Body Weight and Food/Eating-Related Behaviours during COVID-19 Pandemic, Traumatic or Stressful Life Events

Body Weight and Food/Eating-Related Behaviours during COVID-19 Pandemic, Traumatic or Stressful Life Events

Guest Editors

Marios Argyrides
Antonios Dakanalis

MDPI Basel • Beijing • Wuhan • Barcelona • Belgrade • Novi Sad • Cluj • Manchester

Guest Editors
Marios Argyrides
Department of Psychology
Neapolis University Pafos
Paphos
Cyprus

Antonios Dakanalis
Department of Medicine and Surgery
University of Milano Bicocca
Monza
Italy

Editorial Office
MDPI AG
Grosspeteranlage 5
4052 Basel, Switzerland

This is a reprint of the Special Issue, published open access by the journal *Nutrients* (ISSN 2072-6643), freely accessible at: https://www.mdpi.com/journal/nutrients/special_issues/90BSD41K93.

For citation purposes, cite each article independently as indicated on the article page online and as indicated below:

Lastname, A.A.; Lastname, B.B. Article Title. *Journal Name* **Year**, *Volume Number*, Page Range.

ISBN 978-3-7258-6248-1 (Hbk)
ISBN 978-3-7258-6249-8 (PDF)
https://doi.org/10.3390/books978-3-7258-6249-8

Contents

About the Editors

Marios Argyrides

Marios Argyrides is a Professor of Counseling Psychology and the Head of the Department of Psychology at Neapolis University Pafos (Cyprus). He holds a PhD in Counseling Psychology from Tennessee State University and an MSc in Applied Psychology from the University of South Alabama. His research interests include eating disorders, body image, visible differences, body-positivity interventions, applied psychology, and cross-cultural psychology. He has published extensively on disordered eating behaviors and body image satisfaction in various populations and has led work on psychological interventions aiming to enhance body positivity. His dedication to advancing the psychological science of appearance-related issues and his role in educational leadership make him a valued editor for this Special Issue.

Antonios Dakanalis

Antonios Dakanalis is a Professor and Lead Researcher in Psychiatry, Mental Health Services, and Psychotherapy in the Department of Medicine and Surgery at the University of Milano-Bicocca (Italy). His wide-ranging interests cover eating disorders, personality disorders, anxiety and depression, longitudinal research, risk factors and prevention, evidence-based therapy, experimental psychiatry, trauma, neuroscience, body image and weight, adolescence, e-health, and virtual reality interventions. He has pioneered the use of virtual reality and digital technologies for body-image and weight-related psychopathology and has contributed high-impact work on the nosology, assessment, and treatment of binge eating and other eating disorders. His international profile and multidisciplinary orientation further strengthen the editorial team of the Special Issue.

Preface

This Reprint brings together interdisciplinary research on how the COVID-19 pandemic and other stressful life events shaped body weight, eating behaviors, and mental health. Its purpose is to present updated evidence on the interplay between lifestyle, stress, and health outcomes, offering insights valuable to clinicians, researchers, educators, and public health professionals. Motivated by the need to understand the lasting effects of global disruption, this volume aims to support future prevention and intervention efforts across diverse populations.

Marios Argyrides and Antonios Dakanalis
Guest Editors

nutrients

MDPI

Editorial

Body Weight and Food/Eating-Related Behaviors During the COVID-19 Pandemic or Other Traumatic or Stressful Life Events

Marios Argyrides [1] and Antonios Dakanalis [2,3,*]

1 Department of Psychology, Neapolis University Pafos, Paphos 8042, Cyprus; m.argyrides.1@nup.ac.cy
2 Department of Mental Health, Fondazione IRCSS San Gerardo dei Tintori, Via G.B. Pergolesi 33, 20900 Monza, Italy
3 Department of Medicine and Surgery, University of Milano Bicocca, Via Cadore 38, 20900 Monza, Italy
* Correspondence: antonios.dakanalis@unimib.it

1. Introduction

The COVID-19 pandemic has been a pivotal event, reshaping many aspects of daily life and public health across the globe. As the world grapples with the ongoing repercussions of the pandemic, research has increasingly focused on how COVID-19 and its related restrictions have impacted various aspects of human behavior and health. This Special Issue includes nine significant studies exploring the pandemic's effects on lifestyle factors, eating behaviors, and mental health. The findings presented provide a comprehensive view of how the pandemic has influenced public health and underscore the importance of adopting informed strategies to address these impacts.

2. The Interplay Between Lifestyle Factors and COVID-19 Susceptibility

One critical area of research has involved understanding how lifestyle factors affect susceptibility to COVID-19. In a study by Akbar et al. [1] data from the Qatar Biobank was used to explore associations between lifestyle factors and COVID-19 susceptibility. This large-scale analysis involving 10,000 participants found that certain lifestyle choices, such as smoking status and Vitamin D supplementation, were linked to varying risks of contracting COVID-19. Specifically, current and ex-smokers had lower odds of contracting the virus compared to non-smokers, while Vitamin D supplementation was associated with an 18% reduction in the likelihood of infection. Conversely, obesity and adherence to a modern dietary pattern high in saturated fats and refined carbohydrates were positively associated with increased COVID-19 risk. These findings highlight the potential role of lifestyle modifications in mitigating the risk of infectious diseases, emphasizing the importance of public health interventions that promote healthier habits.

3. Pandemic Restrictions and Behavioral Changes in Specific Populations

The pandemic's restrictions have led to significant changes in individual behaviors, particularly among vulnerable populations. Al-Haifi et al. [2] focused on Kuwaiti college students, revealing that pandemic-related restrictions had notable effects on their anthropometric measurements and lifestyle behaviors. The study found that physical activity levels and dietary habits changed significantly, contributing to changes in body weight. These findings are indicative of the impact of the pandemic on young adults and the need for targeted interventions to promote healthy lifestyle choices in this population.

Another study in this Special Issue [3] examined the effects of confinement on individuals at high risk for type 2 diabetes. Findings showed that the lockdown had a detrimental impact on participants' health-related habits, including reduced physical activity and perceived worse diet quality. However, decreased weight and BMI were also observed, and adherence to the Mediterranean diet was improved. This mixed outcome highlights the complex nature of behavioral responses to confinement and suggests that while some individuals may adopt healthier eating patterns, others may respond in more negative ways, negatively impacting their health.

4. The Psychological Impact of Home Confinement and Stress

In another study in this Special Issue, Aljaadi et al. [4] assessed the effects of psychological stress resulting from home confinement and pandemic restrictions on health behaviors in Saudi Arabia and found that higher stress levels were associated with irregular eating patterns and weight gain. The study also noted that stress was linked to poorer sleep quality and disrupted eating routines, further compounding the adverse effects of confinement on health. Similarly, Foroughi et al. [5] conducted a 15-year follow-up study on women in the community, revealing an increase in eating disorder symptoms during the pandemic. The disruption of normal eating patterns, coupled with reduced social support and exercise restrictions, contributed to an increased eating disorder symptomatology. These studies emphasize the importance of addressing mental health and stress-related factors to mitigate their impact on eating behaviors and overall well-being.

5. Social Interactions and Eating Disorders

In another study featured in this Special Issue, Meneguzzo et al. [6] investigated how virtual social inclusion or ostracism affected emotions and eating psychopathology in patients with various eating disorders. Their study revealed that social exclusion negatively impacted psychological well-being in patients with anorexia nervosa. In contrast, those with bulimia nervosa experienced more negative effects from overinclusion. Patients with binge-eating disorder showed a reduction in specific negative emotions after overinclusion. These findings highlight the complex interplay between social dynamics and eating disorders, suggesting that both social inclusion and exclusion can significantly influence eating behaviors and emotional states.

6. The Impact of Chronic Stress on Childhood Obesity

In another study in this Special Issue, Piatkowska-Chmiel et al. [7] explored how chronic stress related to COVID-19 affected eating behaviors and obesity risk in children and adolescents. Their study found that chronic stress contributed to unhealthy eating patterns, weight gain, and disrupted hormonal balance, including reduced leptin and increased ghrelin levels. These hormonal changes, along with stress-induced behavioral changes, were linked to increased hunger and uncontrolled snacking. The study emphasizes the multifaceted nature of stress-related obesity and underscores the need for comprehensive preventive and intervention strategies to address these issues in children and adolescents.

7. Public Interest in Energy Labeling and Dietary Choices

Another aspect of pandemic-related behavioral change addressed by this Special Issue is the public's interest in energy labeling on restaurant menus. Alkhaldy et al. [8] examined how the COVID-19 pandemic influenced interest in energy labeling and its impact on food choices. The study found that while there was some awareness of energy labeling during the pandemic, the level of interest did not significantly change compared to pre-pandemic times. This suggests that public engagement with energy labeling remains relatively low despite the increased focus on health. The findings point to a potential area for improvement in public health strategies, particularly in enhancing consumer education and engagement with nutritional information.

8. Systematic Review of Binge-Eating Disorder During the Pandemic

Lastly, Caldiroli et al. [9] conducted a systematic review to synthesize the impact of the COVID-19 pandemic on binge-eating disorder (BED). Their review found an increase in new BED diagnoses and in symptoms among individuals with pre-existing BED. The study highlights the vulnerability of BED patients to pandemic-related stressors and social isolation, suggesting a need for continued monitoring and targeted interventions. The review also notes the limitations of the existing literature, including the variability in study quality and heterogeneity in pandemic phases and populations.

9. Conclusions: A Call for Holistic Public Health Approaches

The findings presented in this Special Issue offer a better understanding of the diverse impact of the COVID-19 pandemic on lifestyle, eating behaviors, and overall health. From the effects of lifestyle factors on COVID-19 susceptibility to the psychological and behavioral changes resulting from pandemic restrictions, this Special Issue provides valuable insights into the challenges faced by different populations. The findings highlight the importance of adopting a holistic approach to public health that addresses both physical and mental health, particularly in situations of prolonged stress and social disruption.

As society continues to navigate the aftermath of the pandemic, these insights will be crucial in shaping effective public health strategies and interventions. By understanding the complex interplay between lifestyle factors, mental health, and eating behaviors, we can better address the needs of vulnerable populations and promote healthier behaviors in a post-pandemic world.

Conflicts of Interest: The authors declare no conflict of interest.

References

1. Akbar, Z.; Kunhipurayil, H.H.; Saliba, J.; Ahmad, J.; Al-Mansoori, L.; Al-Khatib, H.A.; Al Thani, A.A.; Shi, Z.; Shaito, A.A. The Association between Lifestyle Factors and COVID-19: Findings from Qatar Biobank. *Nutrients* **2024**, *16*, 1037. [CrossRef] [PubMed]
2. Al-Haifi, A.R.; Bumaryoum, N.Y.; Al-Awadhi, B.A.; Alammar, F.A.; Alkhalaf, B.N.; Al-Hazzaa, H.M. How COVID-19 Pandemic Restrictions Affected Kuwaiti College Students' Anthropometry, Lifestyle Behaviors, and Dietary Habits. *Nutrients* **2023**, *15*, 4773. [CrossRef]
3. Ochoa Esteban, D.; Martin-Ridaura, C.; Berlinches-Zapero, C.; Ruiz-Fernández, D.; Sanz-Martín, V.; Gavira-Izquierdo, R.; Muñoz-Haba, A.; March, S.; Ceinos-Arcones, M. Impact of COVID-19 Confinement on the Health-Related Habits of People at High Risk of Type 2 Diabetes. *Nutrients* **2023**, *15*, 841. [CrossRef] [PubMed]
4. Aljaadi, A.M.; Bogis, R.J.; Alruhili, N.A.; Alharbi, S.O.; Noorwali, E.A. Stress during Home Confinement Is Associated with Eating Misalignment among Adults during COVID-19 Lockdown. *Nutrients* **2023**, *15*, 4018. [CrossRef] [PubMed]
5. Foroughi, N.; Hay, P.; Mannan, H. Community Women's Lifestyle and Eating Disorders in the Era of COVID-19 Pandemic: A 15-Year Follow-Up Study. *Nutrients* **2023**, *15*, 1676. [CrossRef]
6. Meneguzzo, P.; Meregalli, V.; Collantoni, E.; Cardi, V.; Tenconi, E.; Favaro, A. Virtual Rejection and Overinclusion in Eating Disorders: An Experimental Investigation of the Impact on Emotions, Stress Perception, and Food Attitudes. *Nutrients* **2023**, *15*, 1021. [CrossRef]
7. Piatkowska-Chmiel, I.; Krawiec, P.; Zietara, K.J.; Pawłowski, P.; Samardakiewicz, M.; Pac-Kozuchowska, E.; Herbet, M. The Impact of Chronic Stress Related to COVID-19 on Eating Behaviors and the Risk of Obesity in Children and Adolescents. *Nutrients* **2024**, *16*, 54. [CrossRef] [PubMed]
8. Alkhaldy, A.A.; Alhumaidan, O.A.; Alkhunein, S.M.; Alkhalaf, M.M.; Bookari, K.A.; Arrish, J.M. The Impact of the COVID-19 Pandemic on Public Interest in the Energy Labelling on Restaurant Menus. *Nutrients* **2023**, *15*, 466. [CrossRef]
9. Caldiroli, A.; La Tegola, D.; Manzo, F.; Scalia, A.; Affaticati, L.M.; Capuzzi, E.; Colmegna, F.; Argyrides, M.; Giaginis, C.; Mendolicchio, L.; et al. The Impact of the COVID-19 Pandemic on Binge Eating Disorder: A Systematic Review. *Nutrients* **2023**, *15*, 3777. [CrossRef] [PubMed]

nutrients

MDPI

Article

The Association between Lifestyle Factors and COVID-19: Findings from Qatar Biobank

Zoha Akbar [1], Hasna H. Kunhipurayil [2], Jessica Saliba [3,4], Jamil Ahmad [5], Layla Al-Mansoori [2,6], Hebah A. Al-Khatib [2], Asmaa A. Al Thani [2,6], Zumin Shi [1,*] and Abdullah A. Shaito [2,6,7,*]

1 Department of Human Nutrition, College of Health Sciences, QU Health, Qatar University, Doha P.O. Box 2713, Qatar
2 Biomedical Research Center, Qatar University, Doha P.O. Box 2713, Qatar; h.alkhatib@qu.edu.qa (H.A.A.-K.)
3 Department of Public Health, Faculty of Health Sciences, University of Balamand, Beirut P.O. Box 100, Lebanon
4 Department of Biology, Faculty of Sciences, Lebanese University, Beirut P.O. Box 90656, Lebanon
5 Hamad Medical Corporation, Doha P.O. Box 3050, Qatar
6 Department of Biomedical Sciences, College of Health Sciences, QU Health, Qatar University, Doha P.O. Box 2713, Qatar
7 Department of Basic Medical Sciences, College of Medicine, QU Health, Qatar University, Doha P.O. Box 2713, Qatar
* Correspondence: zumin@qu.edu.qa (Z.S.); abdshaito@qu.edu.qa (A.A.S.)

Abstract: Coronavirus Disease 2019 (COVID-19) manifestations range from mild to severe life-threatening symptoms, including death. COVID-19 susceptibility has been associated with various factors, but studies in Qatar are limited. The objective of this study was to investigate the correlation between COVID-19 susceptibility and various sociodemographic and lifestyle factors, including age, gender, body mass index, smoking status, education level, dietary patterns, supplement usage, physical activity, a history of bariatric surgery, diabetes, and hypertension. We utilized logistic regression to analyze these associations, using the data of 10,000 adult participants, aged from 18 to 79, from Qatar Biobank. In total, 10.5% (n = 1045) of the participants had COVID-19. Compared to non-smokers, current and ex-smokers had lower odds of having COVID-19 (odds ratio [OR] = 0.55; 95% CI: 0.44–0.68 and OR = 0.70; 95% CI: 0.57–0.86, respectively). Vitamin D supplement use was associated with an 18% reduction in the likelihood of contracting COVID-19 (OR = 0.82; 95% CI: 0.69–0.97). Obesity (BMI $\geq$ 30 kg/m^2), a history of bariatric surgery, and higher adherence to the modern dietary pattern—characterized by the consumption of foods high in saturated fat and refined carbohydrates—were positively associated with COVID-19. Our findings indicate that adopting a healthy lifestyle may be helpful in the prevention of COVID-19 infection.

Keywords: COVID-19; smoking; vitamin D; obesity; bariatric surgery; dietary patterns

1. Introduction

Coronavirus Disease 2019 (COVID-19), caused by Severe Acute Respiratory Syndrome Coronavirus 2 (SARS-CoV-2), first emerged in late 2019 and was declared a global pandemic in March 2020. As of early November 2023, the World Health Organization (WHO) had recorded 771 million confirmed COVID-19 cases and over 6.97 million deaths worldwide [1]. COVID-19 exhibits a significant range of variations in its presentation among individuals affected by the virus. Some may remain asymptomatic throughout the course of infection, while others have symptoms that range from a mild upper respiratory tract infection to life-threatening pneumonia [2]. Those with a poor disease prognosis often suffer from Acute Respiratory Distress Syndrome (ARDS) or immune-mediated lung injury, which can be fatal [3]. COVID-19 disease severity may be attributed to immune dysregulation, leading to differences in immune response, stemming from but not limited to the risk factors of older age and comorbidities [4]. While anti-viral therapies and vaccines provide some protection against disease severity, emerging research suggests that lifestyle factors, such as dietary patterns and the use of dietary supplements, may play a role in mitigating disease severity through their anti-inflammatory and immune-modulating properties [5,6].

Numerous studies have demonstrated the association of sociodemographic and behavioral covariates with a higher risk of the acquisition of COVID-19 infection [7–9]. Age, gender, ethnicity, education, socioeconomic status, chronic diseases (e.g., hypertension, obesity, and diabetes), dietary patterns, and supplements have been associated with COVID-19 [10–12]. Among dietary supplements, vitamin D is the most extensively researched, where numerous studies have examined the association between vitamin D status and COVID-19 outcomes [6]. Associations of comorbid medical conditions with the acquisition of SARS-CoV-2 infection are less defined. However, diabetes and obesity are more robustly associated with SARS-CoV-2 infection [7,13–16].

The persistent and widespread effects of the COVID-19 pandemic, coupled with the emergence of new variants of the virus, emphasize the importance of further exploring the numerous factors that influence COVID-19 susceptibility and outcomes, especially in the Qatari population, which is plagued with a high burden of obesity and prevalence of bariatric surgery. Globally, Qatar holds the highest rate of bariatric surgery, with a prevalence as high as 12% [17].

Previous research has not explored the connections between lifestyle, sociodemographic factors, including a history of bariatric surgery, and susceptibility to COVID-19 in Qatar. This study aims to evaluate the relationship between COVID-19 and various lifestyle and demographic factors among adults, utilizing data from the Qatar Biobank (QBB). Additionally, we investigated whether a history of bariatric surgery is associated with COVID-19.

2. Materials and Methods

2.1. Design of Study and Sample Collection

This retrospective cross-sectional study analyzed the data of 10,000 participants in the QBB study. QBB is an ongoing population-based prospective cohort registry, which was established in 2012 by the Qatar Foundation and the Supreme Council of Health in collaboration with Hamad Medical Corporation (HMC). QBB aims to collect extensive biological, clinical, and lifestyle information on 60,000 men and women [18].

QBB participant recruitment takes place through the internet, social media, or family and friends. Upon recruitment, the participants are invited for a health examination at the QBB facility at HMC. The examination involves a self-administered questionnaire and a nurse interview. Sociodemographic data, lifestyle factors, and dietary habits are collected through questionnaires. In addition, interviews conducted by trained research nurses collect information on medical history, family history of disease, and use of medications. The participants undergo physical assessments, including anthropometric measurements such as weight and height *via* a Seca stadiometer, body composition, along with assessments of heart and lung function, blood pressure, grip strength, a treadmill walking test, and a retinal eye test. Saliva, urine, and 60 mL blood samples are also collected and analyzed for a total of 66 biomarkers at HMC [18].

In this study, 10,000 men and women, between the ages of 18 and 79 years, were randomly selected from 20,328 participants recruited by the QBB study. The inclusion criteria were participants > 18 years old and the availability of complete information on the exposure and outcome variables. Participants < 18 years of age (n = 222) and participants with missing data on COVID-19 and the exposure variables (n = 0) were excluded. Those with complete information on the exposure and outcome variables were eligible for inclusion in this study. Figure 1 shows the process of the selection of the participants from the QBB dataset.

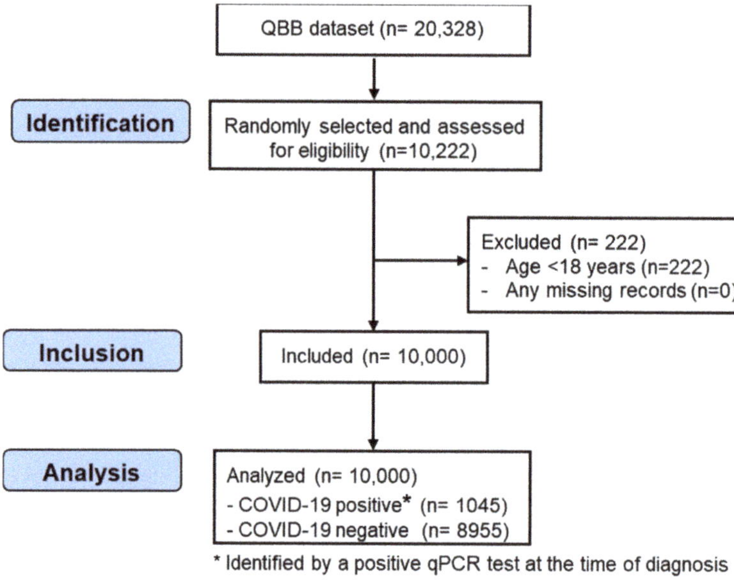

Figure 1. STROBE flow chart illustrating the identification of the included QBB participants. QBB: Qatar Biobank; qPCR: quantitative Polymerase Chain Reaction.

2.2. Outcome Variable: COVID-19 Infection

The status of COVID-19 infection was ascertained through laboratory diagnosis (positive qPCR test) at the time of infection [19].

This diagnosis was performed by QBB's clinical staff, and the COVID-19-positive QBB participants were included in the Qatar COVID-19 biorepository—a national initiative designed to facilitate extensive biomedical research to address the demand for high-quality clinical data on COVID-19 in Qatar. Patients diagnosed with COVID-19 were enrolled during their illness period from three main public hospitals in Qatar. This recruitment took place over a span of 7 months, from March to September 2020. QBB and the Qatar COVID-19 biorepository are linked. In fact, the QBB registry contains two types of COVID-19-positive participants: (1) participants already recruited to the QBB study, and (2) participants who are diagnosed with COVID-19 and have not already ben recruited to the QBB study. The latter group was recruited to the QBB study at the time of positive COVID-19 diagnosis and underwent the same interviews and medical assessments as the existing QBB participants.

2.3. Exposure Variables: Sociodemographic and Lifestyle Factors

Various sociodemographic and lifestyle factors were assessed, including age, gender, education level (lower: up to secondary school, intermediate: technical or professional school, but below university level, and higher: university graduate or higher), smoking status (non-smokers, ex-smokers, and current smokers), and physical activity level (metabolic equivalent of task [MET] hours/week, recoded as tertiles), which were collected using questionnaires filled through face-to-face interviews conducted at the QBB clinic by professional nurses. Data were also obtained on history of diabetes and hypertension. Diabetes was defined as a self-reported doctor diagnosis, HbA1c $\geq$ 6.5%, Random Blood Glucose (RBG) $\geq$ 11.1 mmol/L, or Fasting Blood Glucose (FBG) $\geq$ 7 mmol/L. Hypertension was defined as a self-reported doctor diagnosis, Systolic Blood Pressure $\geq$ 140 mmHg, or Diastolic Blood Pressure $\geq$ 90 mmHg. Body Mass Index (BMI) was assessed and categorized as normal (18.5–24.9 kg/m^2), overweight (25–29.5 kg/m^2), and obese ($\geq$30 kg/m^2). The dietary habits of the QBB participants were evaluated through a self-administered computer-based food frequency questionnaire (FFQ), comprising 102 food items to assess the frequency of their consumption of various foods and beverages over the past year [18]. Although derived from the European Prospective Investigation into Cancer and Nutrition (EPIC) study, this particular FFQ has not undergone validation among the Qatari population. Nonetheless, the food items in this questionnaire closely align with those of a recently validated FFQ used in Qatar [20]. A history of bariatric surgery and dietary supplement use (multivitamin/minerals, iron, calcium, folic acid, vitamin D, vitamin C, vitamin B, and others) were assessed via a nurse-administered interview, while serum vitamin D levels (ng/mL) were measured.

2.4. Statistical Analysis

Quantitative variables were described as mean ± standard deviation (SD) and qualitative (categorical) variables were described as frequencies (counts) and percentages. A student *t*-test and Chi-square test were used to assess the differences in the continuous variables and categorical variables in terms of COVID-19 infection status, respectively. Dietary patterns were identified using a factor analysis. The number of dietary factors to retain was determined by Eigen values > 1, a scree plot, and the interpretability of the identified patterns. Logistic regression was used to examine the association between the exposure variables and COVID-19 infection. Two models were used for the primary analysis and computation of the OR with a 95% confidence interval (CI). Model 1 was adjusted for age and gender, and model 2 was further adjusted for education, smoking, physical activity, BMI categories, diabetes, hypertension, and any supplement use. Associations between COVID-19 and the consumption of various dietary supplements were also assessed by using logistic regression across three multivariable models. Model 3 was similar to model 2, with the exception of an adjustment made for bariatric surgery, as opposed to BMI in Model 2. Subgroup analyses were conducted, and multiplicative interactions between gender, BMI levels, hypertension, diabetes, smoking, and serum vitamin D levels combined with vitamin D supplement use were tested. All the analyses were conducted using STATA software (Version 17, Stata Corporation, College Station, TX, USA) [21], and p-values < 0.05 (2-tailed) were considered to be statistically significant.

2.5. Ethical Approval

The Institutional Review Boards at QBB (Approval number: EX-2021-QF-QBB-RES-ACC-00065-0181) and Qatar University (QU-IRB 1779-E/22) ethically approved the study. The study sample was obtained from QBB in accordance with the principles outlined in the Declaration of Helsinki, which is designed to safeguard the well-being of human subjects involved in Biomedical Research [22]. QBB obtains informed consent from the participants enrolled in the study, while maintaining anonymity and the non-disclosure of personal identity.

3. Results

3.1. Sample Characteristics

The mean age of the participants in this study was 40.3 ± 13.1 years, and 52.2% were females. The overall prevalence of obesity and history of bariatric surgery was 44.2% and 12.1%, respectively. Of the included 10,000 participants, 1045 (10.5%) had COVID-19 (Table 1). A descriptive analysis showed that individuals with COVID-19 infection were more likely to be obese, non-smokers, more physically active, and have higher scores for modern dietary pattern. They were also more likely to have a history of bariatric surgery (14.9%) and have reported a higher use of dietary supplements (61.1%) than those without COVID-19. Among vitamin and mineral supplements, the most reported supplements were multivitamin/minerals (37.4%) and vitamin D (22.6%). Supplementary Figure S1 shows the distribution of serum vitamin D in the study population, where the mean serum vitamin D level was 19.29 (ng/mL). Totals of 16.4% and 20.3% of the participants had hypertension and diabetes, respectively (Table 1).

Table 1. Sample characteristics by COVID-19 infection status.

	Total N = 10,000	No N = 8955	Yes N = 1045	p-Value
Age (years)	40.3 ± 13.1	40.4 ± 13.1	39.6 ± 12.7	0.083
Gender				<0.001
Male	4780 (47.8%)	4351 (48.6%)	429 (41.1%)	
Female	5220 (52.2%)	4604 (51.4%)	616 (58.9%)	
Educational level				0.020
Low	1796 (18.0%)	1594 (17.8%)	202 (19.3%)	
Medium	3040 (30.4%)	2694 (30.1%)	346 (33.1%)	
High	5164 (51.6%)	4667 (52.1%)	497 (47.6%)	
Smoking status				<0.001
None	6468 (64.7%)	5695 (63.6%)	773 (74.0%)	
Smoker	1866 (18.7%)	1735 (19.4%)	131 (12.5%)	
Ex-smoker	1666 (16.7%)	1525 (17.0%)	141 (13.5%)	
Leisure time physical activity (MET hours/week)	24.9 ± 52.5	24.9 ± 51.0	24.3 ± 64.2	0.74
Modern dietary pattern score	0.0 ± 1.00	−0.01 ± 1.0	0.1 ± 1.1	0.001
Prudent dietary pattern score	0.0 ± 1.00	−0.0 ± 1.0	0.02 ± 0.9	0.45
Convenience dietary pattern score	0.0 ± 1.00	−0.01 ± 1.0	0.1 ± 1.1	0.032
Serum vitamin D (ng/mL)	19.3 ± 11.1	19.3 ± 1.1	18.9 ± 10.9	0.22
Hypertension	1638 (16.4%)	1479 (16.5%)	159 (15.2%)	0.28
Diabetes	2026 (20.3%)	1808 (20.3%)	218 (21.0%)	0.60
History of bariatric surgery	1213 (12.1%)	1058 (11.8%)	155 (14.9%)	0.004
BMI categories				0.051
Normal	2112 (21.1%)	1905 (21.3%)	207 (19.8%)	
Overweight	3462 (34.7%)	3123 (34.9%)	339 (32.4%)	
Obese	4417 (44.2%)	3918 (43.8%)	499 (47.8%)	
Dietary supplement use				
Multivitamin/minerals	3565 (35.6%)	3174 (35.4%)	391 (37.4%)	0.21
Calcium	726 (7.3%)	649 (7.2%)	77 (7.4%)	0.89
Folic acid	316 (3.2%)	276 (3.1%)	40 (3.8%)	0.19
Iron	1366 (13.7%)	1208 (13.5%)	158 (15.1%)	0.15
Vitamin B	806 (8.1%)	728 (8.1%)	78 (7.5%)	0.45
Vitamin C	665 (6.7%)	604 (6.7%)	61 (5.8%)	0.27
Vitamin D	2402 (24.0%)	2166 (24.2%)	236 (22.6%)	0.25
Other supplements	3315 (33.1%)	2968 (33.1%)	347 (33.2%)	0.97
Any supplement use	5977 (59.8%)	5339 (59.6%)	638 (61.1%)	0.37

Data are presented as mean ± SD for continuous measures, and *n* (%) for categorical measures. For the dietary patterns, the numbers represent mean dietary pattern scores ± SD with a mean factor score as zero, and an SD of 1 after factor analysis (Supplementary Table S1). Higher dietary mean scores are present in participants with COVID-19, and lower scores in those without COVID-19. MET: metabolic equivalent of task.

Three major dietary patterns were identified, and their factor loadings are shown in Supplementary Table S1. Accordingly, the foods identified in each pattern were named as modern, prudent, and convenience patterns, respectively. The three dietary patterns explained 33.5% of the variation in food intake, which is comparable to other published studies on dietary patterns in the region [23–25]. The modern pattern was characterized by foods high in saturated fat and refined carbohydrates such as fast food, lasagna, mixed dishes of chicken, meat and fish, soft drinks, biryani, desserts, ice cream, chocolate, Asian noodles, zaatar fatayer (thyme and sesame spread on flour dough), croissants, and potatoes. The prudent pattern had high factor loadings of cooked and raw vegetables, fruits, fish, soups and starters, fruit juice, and nuts. The convenience pattern consisted of foods that are ready-to-eat or require minimal preparation, such as yoghurt, cheese, cereal, milkshakes, butter, croissants, zaatar fatayer, and many types of breads (white bread, brown bread, and Arabic bread).

3.2. Sociodemographic and Lifestyle Factors and COVID-19 Infection

Among the sociodemographic and lifestyle factors, a statistically significant association of COVID-19 was found with vitamin D, smoking status, obesity, and the modern dietary pattern (Table 2). Vitamin D supplement usage was associated with an 18% reduced likelihood of COVID-19 (OR = 0.82; 95% CI: 0.69–0.97; p = 0.022) in the fully adjusted model. Compared to non-smokers, current and ex-smokers had 45% and 30% lower odds of having COVID-19, respectively ($p < 0.001$). Using the normal BMI category as a reference group, participants who were obese were 20% more likely to have COVID-19 infection after adjusting for covariates (OR = 1.20; 95% CI: 1.00–1.45; p-value = 0.049). Among the three dietary patterns, the modern dietary pattern was positively associated with COVID-19, and the odds increased by 9% per 1 SD increment in modern dietary pattern score (p = 0.012). A history of bariatric surgery was also associated with 24% higher odds of COVID-19 infection (OR = 1.24; 95% CI: 1.03–1.50). No associations were found between gender, education, physical activity, diabetes, hypertension, other dietary supplements, and other dietary patterns with COVID-19 (Tables 2 and 3).

Table 2. Association between sociodemographic and lifestyle factors with COVID-19 infection (n = 10,000).

	Model 1 OR [95% CI]	***p*-Value**	**Model 2 OR [95% CI]**	***p*-Value**
Vitamin D	0.81 (0.68–0.96)	0.018	0.82 (0.69–0.97)	0.022
Age (years)	0.99 (0.99–1.00)	0.050	0.99 (0.99–1.00)	0.084
Gender				
Male	1.00		1.00	
Female	1.03 (0.88–1.21)	0.715	1.03 (0.88–1.21)	0.704
Education				
Low	1.00		1.00	
Medium	0.95 (0.77–1.17)	0.641	0.94 (0.76–1.17)	0.588
High	0.86 (0.71–1.04)	0.124	0.85 (0.70–1.03)	0.096
Smoking				
Non	1.00		1.00	
Smoker	0.55 (0.44–0.68)	<0.001	0.55 (0.44–0.68)	<0.001
Ex-smoker	0.70 (0.57–0.86)	<0.001	0.70 (0.57–0.86)	<0.001
Leisure time PA (MET hours/week)				
T1	1.00		1.00	
T2	1.02 (0.87–1.20)	0.817	1.03 (0.88–1.21)	0.737
T3	0.94 (0.79–1.11)	0.472	0.93 (0.79–1.10)	0.404
BMI level				
Normal	1.00			
Overweight	1.09 (0.90–1.32)	0.379		
Obese	1.20 (1.00–1.45)	0.049		
Diabetes				
No	1.00		1.00	
Yes	1.09 (0.91–1.31)	0.338	1.10 (0.92–1.32)	0.308
Hypertension				
No	1.00		1.00	
Yes	0.94 (0.76–1.16)	0.547	0.96 (0.78–1.18)	0.693
Modern dietary pattern	1.09 (1.02–1.16)	0.012	1.08 (1.01–1.16)	0.018
Prudent dietary pattern	1.03 (0.97–1.10)	0.344	1.03 (0.97–1.10)	0.307
Convenience dietary pattern	1.05 (0.99–1.12)	0.100	1.05 (0.99–1.12)	0.092
Any supplement use	1.12 (0.96–1.30)	0.155	1.10 (0.95–1.28)	0.205
History of bariatric surgery				
No			1.00	
Yes			1.24 (1.03–1.50)	0.022

All the variables in the model were mutually adjusted. BMI: body mass index; CI: confidence interval; MET: metabolic equivalent of task; OR: odds ratio; PA: physical activity; and T: tertile.

Table 3. Association between dietary supplement use and COVID-19 infection (n = 10,000).

	Model 1 OR (95% CI)	p-Value	Model 2 OR (95% CI)	p-Value	Model 3 OR (95% CI)	p-Value
Any supplement use	1.00 (0.87–1.14)	0.979	1.03 (0.90–1.19)	0.652	1.02 (0.89–1.18)	0.749
Multivitamin/minerals	1.07 (0.94–1.22)	0.322	1.17 (0.99–1.40)	0.067	1.16 (0.97–1.38)	0.096
Calcium	0.96 (0.75–1.22)	0.723	0.97 (0.75–1.25)	0.792	0.96 (0.75–1.24)	0.775
Folic acid	1.13 (0.80–1.59)	0.485	1.18 (0.83–1.66)	0.356	1.17 (0.82–1.65)	0.385
Iron	1.01 (0.84–1.22)	0.886	1.02 (0.84–1.24)	0.869	1.00 (0.82–1.22)	1.000
Other supplements	0.94 (0.81–1.08)	0.357	0.90 (0.76–1.06)	0.197	0.90 (0.76–1.06)	0.216
Vitamin B	0.87 (0.68–1.12)	0.281	0.89 (0.69–1.14)	0.348	0.87 (0.68–1.12)	0.288
Vitamin C	0.84 (0.64–1.10)	0.210	0.86 (0.65–1.14)	0.288	0.86 (0.65–1.13)	0.276
Vitamin D	0.85 (0.73–1.00)	0.046	0.81 (0.68–0.96)	0.018	0.82 (0.69–0.97)	0.022

Model 1 adjusted for age and gender. Model 2 further adjusted for education, smoking, physical activity, BMI, diabetes, hypertension, dietary patterns, and any supplement use. Model 3 was adjusted for bariatric surgery instead of BMI in Model 2. Both Models 2 and 3 adjusted for dietary patterns.

3.3. *Vitamin D and COVID-19 Infection*

Subgroup analyses revealed no significant interactions between vitamin D supplement use and sociodemographic and lifestyle factors in relation to COVID-19. However, an association between vitamin D supplementation and COVID-19 infection was only observed among those with high serum vitamin D levels (OR = 0.71; 95% CI: 0.53–0.95) or those who did not have hypertension (OR = 0.83; 95% CI: 0.70–1.00) (Supplementary Table S2).

Supplementary Figure S1 presents the distribution of serum levels among the participants. The association between quartiles of serum vitamin D and COVID-19 infection is presented in Table 4. In the fully adjusted model, the ORs (95% CI) for COVID-19 infection, from the lowest to highest quartiles, were 1.00 (reference group), 1.19 (1.00–1.43), 1.06 (0.87–1.29), and 0.98 (0.80–1.20) (p = 0.552). We could not detect a significant trend of association between COVID-19 infection and vitamin D serum level quartiles, ranging from high to low serum vitamin D levels.

Table 4. Odds ratio (95% CI) for COVID-19 infection by quartiles of serum vitamin D.

	Quartiles of Serum Vitamin D				
	Q1	Q2	Q3	Q4	p-Value
Model 1	1.00	1.19 (0.99–1.42)	1.04 (0.85–1.26)	0.94 (0.77–1.15)	0.324
Model 2	1.00	1.19 (1.00–1.43)	1.06 (0.87–1.29)	0.98 (0.80–1.20)	0.552

Model 1 adjusted for age and gender. Model 2 further adjusted for education, smoking, physical, BMI, diabetes, hypertension, and any supplement use.

4. Discussion

In this large retrospective cross-sectional analysis of data from 10,000 QBB subjects, we report on the associations between various sociodemographic and lifestyle factors and COVID-19 infection. Dietary patterns were constructed using a factor analysis—a commonly used method in epidemiological research [26]. Smoking (current and past) and vitamin D supplement use were inversely associated with COVID-19 infection, while subjects with obesity, a history of bariatric surgery, and higher modern dietary pattern scores—characterized by foods high in saturated fat and refined carbohydrates (e.g., fast food, soft drinks, and dessert)—had higher odds of COVID-19. No significant associations were found for gender, education, physical activity, diabetes, hypertension, and other dietary supplements or patterns.

4.1. *Smoking Status and COVID-19 Susceptibility*

Lower odds of COVID-19 in the current and former smokers in this sample was an unexpected, albeit not entirely inconsistent, finding. Previous studies have also reported lower rates of smokers among patients diagnosed with COVID-19, compared to non-smokers and former smokers [27–29]. The literature has explored the association between smoking and nicotine with SARS-CoV-2 infection. While an upregulation of angiotensin-converting enzyme 2 (ACE2) receptors, which are used by the virus for cell entry, is linked to nicotine exposure [30,31], it is also possible that nicotine binds ACE2 receptors, thus competing with SARS-CoV-2 binding and reducing the virus competence. In fact, several studies have indicated potential preventive effects of nicotine and/or caffeine in COVID-19 by blocking ACE2 receptor binding and preventing SARS-CoV-2 infection [32, 33]. Furthermore, several reports have demonstrated the binding of DNA fragments of SARS-CoV-2 Spike protein to nicotinic acetylcholine receptors (nAChRs) [34,35]. This observation supports the notion that COVID-19 can dysregulate the nicotinic cholinergic system, and this dysregulation may be involved in the pathophysiology of COVID-19. In this regard, nicotine and other nicotinic cholinergic agonists may protect nAChRs and thus could have a beneficial role in COVID-19 patients [34,35]. Additionally, nicotine has been reported to attenuate the cytokine storm associated with COVID-19 symptoms [34,36,37]. These observations may potentially explain the incidence of COVID-19 among non-smokers compared to smokers in this study. Alternatively, selection bias could play a role, as symptomatic patients, including smokers with coughs, might undergo more frequent testing, potentially skewing the results. However, the opposite—i.e., increased odds of COVID-19 among smokers—has also been reported in the literature, where smoking has been associated with worse outcomes [38,39]. The contribution of smoking to COVID-19 diagnosis may, in fact, depend on the stage of COVID-19 infection, where smoking may have a protective role at the early stages of infection and a pathogenic role at later stages. These notions remain to be investigated in future studies.

4.2. Vitamin D Supplementation and COVID-19 Susceptibility

In this study, COVID-19 was associated with several dietary and nutritional parameters among the Qatari population. Vitamin D supplementation was inversely associated with COVID-19 infection. This is in agreement with many studies that have indicated a protective role of vitamin D against COVID-19 acquisition. For example, vitamin D deficiency has been linked to an increased susceptibility to COVID-19 infection and a higher risk of severe outcomes [40]. A retrospective cohort study found that patients supplemented with vitamin D were at a lower risk of SARS-CoV-2 infection and COVID-19-related mortality [41]. Similarly, a double-blind, parallel randomized controlled trial (RCT), reported that vitamin D supplementation (leading to increased vitamin D serum levels) seemed to protect against SARS-CoV-2 infection, regardless of baseline vitamin D status [42]. In contrast, another RCT reported that individuals with vitamin D insufficiency at baseline who received vitamin D supplementation were not at a lower risk of acute respiratory infections or COVID-19 [43]. In the present study, subgroup analyses showed that vitamin D supplementation was associated with lower odds of COVID-19 in participants with higher serum vitamin D levels. However, no significant interactions between vitamin D supplement use and sociodemographic and lifestyle factors were identified in relation to a positive COVID-19 diagnosis. Vitamin D metabolites play a role in boosting immune responses against respiratory viruses and bacteria [44]. The protective effect of vitamin D against COVID-19 may be explained by the activation of the production of antimicrobial peptides, such as cathelicidin LL-37 and human beta-defensin 2, that can bind to SARS-CoV-2 spike protein, inhibiting its binding to host ACE2 receptors and hindering virus infection [45–47].

4.3. Obesity and History of Bariatric Surgery and COVID-19 Susceptibility

Obesity was expectedly associated with a higher susceptibility to COVID-19 in this study. Obesity is an independent risk factor for COVID-19 and the existing literature has consistently illustrated that individuals with a BMI of ≥ 30 kg/m^2 are more likely to test positive for COVID-19, and experience worse outcomes and higher mortality rates [46].

This is potentially due to obesity-related chronic inflammation, which impairs the immunological responses of the body and its ability to respond to infections [48]. Participants with a history of bariatric surgery had higher odds of contracting COVID-19, although it is worth noting that the observed predisposition to COVID-19 in this study is inconsistent with prior research, where bariatric surgery appeared to be associated with lower mortality rates and a reduced risk of poor outcomes [49–51]. The positive association in this study may be partially explained by the fact that bariatric surgery may result in vitamin and mineral deficiencies and malnutrition, which are known risk factors for various health complications [52,53]. Bariatric surgery is also associated with hypercatabolic and immunosuppressive states in individuals, which may contribute to poorer clinical outcomes after contracting SARS-CoV-2 [54,55]. Alternatively, the observed association may be due to selection bias. Patients with bariatric surgery may have regular follow-ups and are therefore more likely to be routinely tested for COVID-19 compared to surgery-free participants.

4.4. Modern Dietary Pattern

Similar to obesity and bariatric surgery, an unhealthy eating pattern was also associated with higher odds of COVID-19 in this study. A cross-sectional analysis of the impact of dietary patterns on COVID-19 severity in Iran found a positive association of an unhealthy diet with longer hospital stays and worse symptoms in COVID-19 patients [56]. Existing research on the associations of COVID-19 and dietary patterns has primarily reported on plant-based diets, which are associated with a lower COVID-19 infection risk [57] and severity [58,59]. Studies have reported that improving the quality of nutritional intake may reduce the impact of infectious diseases [60,61], partly through arachidonic or linoleic acid that suppress viral replication [62]. Additionally, some B vitamins (e.g., vitamin B6, B12, and folate), vitamins C, D, E, and A, omega-3 fatty acids and amino acids, and the polyphenols found in plant-based foods play crucial roles in the functioning of the immune system and the release of inflammatory mediators [11,62,63]. Unhealthy eating habits, such as the modern eating pattern described in this study, lack these potentially protective food elements, which may explain the observed association between modern eating habits and increased rates of COVID-19. Moreover, consuming high-fat diets triggers the activation of the body's natural immune response and hinders the adaptive immune system, resulting in a chronically inflamed state and compromised protection against viral infections [64]. Personal attitudes and beliefs seem to favor the use of vitamin D and plant-based medicinal products in the treatment or prevention of COVID-19 [65,66], underscoring the role of traditional foodstuffs and dietary recipes in boosting health and curbing infections.

4.5. Strengths and Limitations

To the best of our knowledge, this is the first study to assess the association between COVID-19 and lifestyle and sociodemographic factors in Qatar. This study has some notable strengths. Firstly, the large sample size increases the precision of estimates. Secondly, participants were selected randomly. Owing to the detailed collection process of QBB data, which was started before the COVID-19 pandemic, we were able to adjust for numerous confounding factors; reverse causation is unlikely in the study. Finally, the 102-item FFQ allowed for capturing the most commonly consumed foods in the local population.

However, due to the observational nature of this study and the secondary use of data, no temporality or causality could be established. Additionally, the FFQ questionnaire that was employed was qualitative and did not include portion sizes. This method of diet recall is associated with some measurement errors, notably recall bias.

5. Conclusions

In conclusion, COVID-19 in this Qatari population was inversely associated with smoking (current or past) and vitamin D supplementation. On the contrary, subjects with obesity, a history of bariatric surgery, and higher scores for the modern dietary pattern were more likely to have a positive COVID-19 diagnosis. Overall, this study contributes to the growing body of knowledge on COVID-19 infection and sheds light on some of the factors possibly associated with increased rates of COVID-19, and potentially other respiratory infections. While the associations of lifestyle and sociodemographic factors with COVID-19 are fairly supported by sound evidence, differences in COVID-19 infection rates remain difficult to determine and prove, warranting future investigation. Further research, particularly long-term prospective cohort studies, are needed to assess the temporal relationships between potential risk or protective factors and a COVID-19 diagnosis. Additionally, the high rates of obesity and bariatric surgery in this population, coupled with their connection to COVID-19 infection, highlight the importance of implementing effective public health measures to address the growing obesity epidemic.

Supplementary Materials: The following supporting information can be downloaded at: https://www.mdpi.com/article/10.3390/nu16071037/s1, Figure S1: Distribution of serum levels of vitamin D.; Table S1: Factor loadings of the identified dietary patterns; Table S2: Subgroup analyses of the association between vitamin D supplement use and COVID-19 infection.

Author Contributions: Conceptualization, A.A.S. and Z.S.; investigation, A.A.S., Z.S., Z.A. and J.S.; resources, A.A.S. and Z.S.; writing—original draft preparation, A.A.S., Z.S. and Z.A.; writing—review and editing, All Authors; Supervision, A.A.S., Z.S. and A.A.A.T.; project administration, A.A.S.; funding acquisition, A.A.S. All authors have read and agreed to the published version of the manuscript.

Funding: Qatar National Research Fund (QNRF) grant number RRC02-0807-210,022 to A.A.S. The funding source (QNRF) had no role in study design, in the collection, analysis and interpretation of data, in the writing of the manuscript, or in the decision to submit the article for publication.

Institutional Review Board Statement: This study was approved by Qatar University IRB (QU-IRB 1779-E/22, 11 October 2022) and QBB IRB (EX-2021-QF-QBB-RES-ACC-00065-0181, 25 October 2022).

Informed Consent Statement: Informed consent was obtained from all subjects involved in the study by QBB.

Data Availability Statement: Data are contained within the article.

Conflicts of Interest: The authors declare no conflict of interest.

References

1. WHO Coronavirus (COVID-19) Dashboard. Available online: https://covid19.who.int/ (accessed on 5 November 2023).
2. Wiersinga, W.J.; Rhodes, A.; Cheng, A.C.; Peacock, S.J.; Prescott, H.C. Pathophysiology, Transmission, Diagnosis, and Treatment of Coronavirus Disease 2019 (COVID-19): A Review. *JAMA* **2020**, *324*, 782. [CrossRef] [PubMed]
3. Aslan, A.; Aslan, C.; Zolbanin, N.M.; Jafari, R. Acute Respiratory Distress Syndrome in COVID-19: Possible Mechanisms and Therapeutic Management. *Pneumonia* **2021**, *13*, 14. [CrossRef] [PubMed]
4. Brodin, P. Immune Determinants of COVID-19 Disease Presentation and Severity. *Nat. Med.* **2021**, *27*, 28–33. [CrossRef] [PubMed]
5. Adams, K.K.; Baker, W.L.; Sobieraj, D.M. Myth Busters: Dietary Supplements and COVID-19. *Ann. Pharmacother.* **2020**, *54*, 820–826. [CrossRef] [PubMed]
6. Mitchell, F. Vitamin-D and COVID-19: Do Deficient Risk a Poorer Outcome? *Lancet Diabetes Endocrinol.* **2020**, *8*, 570. [CrossRef] [PubMed]
7. Theodore, D.A.; Branche, A.R.; Zhang, L.; Graciaa, D.S.; Choudhary, M.; Hatlen, T.J.; Osman, R.; Babu, T.M.; Robinson, S.T.; Gilbert, P.B.; et al. Clinical and Demographic Factors Associated with COVID-19, Severe COVID-19, and SARS-CoV-2 Infection in Adults: A Secondary Cross-Protocol Analysis of 4 Randomized Clinical Trials. *JAMA Netw. Open* **2023**, *6*, e2323349. [CrossRef] [PubMed]
8. Rodriguez-Diaz, C.E.; Guilamo-Ramos, V.; Mena, L.; Hall, E.; Honermann, B.; Crowley, J.S.; Baral, S.; Prado, G.J.; Marzan-Rodriguez, M.; Beyrer, C.; et al. Risk for COVID-19 Infection and Death among Latinos in the United States: Examining Heterogeneity in Transmission Dynamics. *Ann. Epidemiol.* **2020**, *52*, 46. [CrossRef]
9. Holt, H.; Talaei, M.; Greenig, M.; Zenner, D.; Symons, J.; Relton, C.; Young, K.S.; Davies, M.R.; Thompson, K.N.; Ashman, J.; et al. Risk Factors for Developing COVID-19: A Population-Based Longitudinal Study (COVIDENCE UK). *Thorax* **2022**, *77*, 900–912. [CrossRef]
10. Kaeuffer, C.; Le Hyaric, C.; Fabacher, T.; Mootien, J.; Dervieux, B.; Ruch, Y.; Hugerot, A.; Zhu, Y.-J.; Pointurier, V.; Clere-Jehl, R.; et al. Clinical Characteristics and Risk Factors Associated with Severe COVID-19: Prospective Analysis of 1045 Hospitalised Cases in North-Eastern France, March 2020. *Eurosurveillance* **2020**, *25*, 2000895. [CrossRef]

11. Louca, P.; Murray, B.; Klaser, K.; Graham, M.S.; Mazidi, M.; Leeming, E.R.; Thompson, E.; Bowyer, R.; Drew, D.A.; Nguyen, L.H.; et al. Modest Effects of Dietary Supplements during the COVID-19 Pandemic: Insights from 445 850 Users of the COVID-19 Symptom Study App. *BMJ Nutr. Prev. Health* **2021**, *4*, 149–157. [CrossRef]
12. Rahmati, M.; Fatemi, R.; Yon, D.K.; Lee, S.W.; Koyanagi, A.; Shin, J.I.; Smith, L. The Effect of Adherence to High-quality Dietary Pattern on COVID-19 Outcomes: A Systematic Review and Meta-analysis. *J. Med. Virol.* **2023**, *95*, e28298. [CrossRef] [PubMed]
13. Ho, J.S.Y.; Fernando, D.I.; Chan, M.Y.; Sia, C.-H. Obesity in COVID-19: A Systematic Review and Meta-Analysis. *Ann. Acad. Med. Singapore* **2020**, *49*, 996–1008. [CrossRef] [PubMed]
14. Najafabadi, B.T.; Rayner, D.G.; Shokraee, K.; Shokraie, K.; Panahi, P.; Rastgou, P.; Seirafianpour, F.; Landi, F.M.; Alinia, P.; Parnianfard, N.; et al. Obesity as an Independent Risk Factor for COVID-19 Severity and Mortality. *Cochrane Database Syst. Rev.* **2023**, *5*, CD015201.
15. Wander, P.L.; Lowy, E.; Beste, L.A.; Tulloch-Palomino, L.; Korpak, A.; Peterson, A.C.; Kahn, S.E.; Boyko, E.J. The Incidence of Diabetes among 2,808,106 Veterans with and without Recent SARS-CoV-2 Infection. *Diabetes Care* **2022**, *45*, 782. [CrossRef] [PubMed]
16. Kumar, S.; Behl, T.; Sachdeva, M.; Sehgal, A.; Kumari, S.; Kumar, A.; Kaur, G.; Yadav, H.N.; Bungau, S. Implicating the Effect of Ketogenic Diet as a Preventive Measure to Obesity and Diabetes Mellitus. *Life Sci.* **2021**, *264*, 118661. [CrossRef] [PubMed]
17. *Annual Report 2022–2023;* Qatar Biobank: Doha, Qatar, 2023.
18. Al Thani, A.; Fthenou, E.; Paparrodopoulos, S.; Al Marri, A.; Shi, Z.; Qafoud, F.; Afifi, N. Qatar Biobank Cohort Study: Study Design and First Results. *Am. J. Epidemiol.* **2019**, *188*, 1420–1433. [CrossRef] [PubMed]
19. Afifi, N.; Fthenou, E.; El Deeb, M.; Frenneaux, M.; Al Thani, A.; Al Khal, A.L.; Biobank, Q.J. Qatar Biobank: COVID-19 Biorepository Project. *Emerg. Med. Trauma Acute Care* **2021**. [CrossRef]
20. Bawadi, H.; Akasheh, R.T.; Kerkadi, A.; Haydar, S.; Tayyem, R.; Shi, Z. Validity and Reproducibility of a Food Frequency Questionnaire to Assess Macro and Micro-Nutrient Intake among a Convenience Cohort of Healthy Adult Qataris. *Nutrients* **2021**, *13*, 2002. [CrossRef]
21. StataCorp. *Stata Statistical Software: Release 17;* StataCorp LLC: College Station, TX, USA, 2023.
22. WMA Declaration of Helsinki–Ethical Principles for Medical Research Involving Human Subjects. Available online: https://www.wma.net/policies-post/wma-declaration-of-helsinki-ethical-principles-for-medical-research-involving-human-subjects/ (accessed on 27 November 2023).
23. Arab, A.; Rafie, N.; Hadi, A.; Khorvash, F.; Heidari, Z.; Askari, G. Empirically Derived Dietary Patterns and Their Association with Mental Health: A Cross-Sectional Sample of Iranian Migraine Patients (2019–2020). *BMC Nutr.* **2022**, *8*, 28. [CrossRef]
24. Fituri, S.; Shi, Z. Association between Dietary Patterns and Cognitive Function among Qatari Adults: A Cross-Sectional Analysis of the Qatar Biobank Study. *Nutrients* **2023**, *15*, 4053. [CrossRef]
25. Tayyem, R.; Al-Qawasmeh, R.; Khawaja, N. Dietary Patterns and Nutrients Associated with Metabolic Syndrome. *Nutr. Food Sci.* **2019**, *50*, 253–268. [CrossRef]
26. Zhao, J.; Li, Z.; Gao, Q.; Zhao, H.; Chen, S.; Huang, L.; Wang, W.; Wang, T. A Review of Statistical Methods for Dietary Pattern Analysis. *Nutr. J.* **2021**, *20*, 37. [CrossRef] [PubMed]
27. de Lusignan, S.; Dorward, J.; Correa, A.; Jones, N.; Akinyemi, O.; Amirthalingam, G.; Andrews, N.; Byford, R.; Dabrera, G.; Elliot, A.; et al. Risk Factors for SARS-CoV-2 among Patients in the Oxford Royal College of General Practitioners Research and Surveillance Centre Primary Care Network: A Cross-Sectional Study. *Lancet Infect. Dis.* **2020**, *20*, 1034. [CrossRef] [PubMed]
28. Paleiron, N.; Mayet, A.; Marbac, V.; Perisse, A.; Barazzutti, H.; Brocq, F.-X.; Janvier, F.; Dautzenberg, B.; Bylicki, O. Impact of Tobacco Smoking on the Risk of COVID-19: A Large Scale Retrospective Cohort Study. *Nicotine Tob. Res.* **2021**, *23*, 1398. [CrossRef] [PubMed]
29. Simons, D.; Shahab, L.; Brown, J.; Perski, O. The Association of Smoking Status with SARS-CoV-2 Infection, Hospitalization and Mortality from COVID-19: A Living Rapid Evidence Review with Bayesian Meta-Analyses (Version 7). *Addiction* **2021**, *116*, 1319. [CrossRef] [PubMed]
30. Maggi, F.; Rosellini, A.; Spezia, P.G.; Focosi, D.; Macera, L.; Lai, M.; Pistello, M.; de Iure, A.; Tomino, C.; Bonassi, S.; et al. Nicotine Upregulates ACE2 Expression and Increases Competence for SARS-CoV-2 in Human Pneumocytes. *ERJ Open Res.* **2021**, *7*, 00713. [CrossRef] [PubMed]
31. Salehi, Z.; Ghoochani, B.F.N.M.; Nourian, Y.H.; Jamalkandi, S.A.; Ghanei, M. The Controversial Effect of Smoking and Nicotine in SARS-CoV-2 Infection. *Allergy Asthma Clin. Immunol.* **2023**, *19*, 49. [CrossRef]
32. Tindle, H.A.; Newhouse, P.A.; Freiberg, M.S. Beyond Smoking Cessation: Investigating Medicinal Nicotine to Prevent and Treat COVID-19. *Nicotine Tob. Res.* **2020**, *22*, 1669. [CrossRef]
33. Mohammadi, S.; Heidarizadeh, M.; Entesari, M.; Esmailpour, A.; Esmailpour, M.; Moradi, R.; Sakhaee, N.; Doustkhah, E. In Silico Investigation on the Inhibiting Role of Nicotine/Caffeine by Blocking the S Protein of SARS-CoV-2 versus ACE2 Receptor. *Microorganisms* **2020**, *8*, 1600. [CrossRef]
34. Farsalinos, K.; Eliopoulos, E.; Leonidas, D.D.; Papadopoulos, G.E.; Tzartos, S.; Poulas, K. Nicotinic Cholinergic System and COVID-19: In Silico Identification of an Interaction between SARS-CoV-2 and Nicotinic Receptors with Potential Therapeutic Targeting Implications. *Int. J. Mol. Sci.* **2020**, *21*, 5807. [CrossRef]
35. Leitzke, M. Is the Post-COVID-19 Syndrome a Severe Impairment of Acetylcholine-Orchestrated Neuromodulation That Responds to Nicotine Administration? *Bioelectron. Med.* **2023**, *9*, 2. [CrossRef] [PubMed]

36. Mehta, P.; McAuley, D.F.; Brown, M.; Sanchez, E.; Tattersall, R.S.; Manson, J.J. COVID-19: Consider Cytokine Storm Syndromes and Immunosuppression. *Lancet* **2020**, *395*, 1033–1034. [CrossRef] [PubMed]
37. Tracey, K.J. The Inflammatory Reflex. *Nature* **2002**, *420*, 853. [CrossRef] [PubMed]
38. Clift, A.K.; von Ende, A.; Tan, P.S.; Sallis, H.M.; Lindson, N.; Coupland, C.A.C.; Munafò, M.R.; Aveyard, P.; Hippisley-Cox, J.; Hopewell, J.C. Smoking and COVID-19 Outcomes: An Observational and Mendelian Randomisation Study Using the UK Biobank Cohort. *Thorax* **2022**, *77*, 65–73. [CrossRef] [PubMed]
39. Gallus, S.; Scala, M.; Possenti, I.; Jarach, C.M.; Clancy, L.; Fernandez, E.; Gorini, G.; Carreras, G.; Malevolti, M.C.; Commar, A.; et al. The Role of Smoking in COVID-19 Progression: A Comprehensive Meta-Analysis. *Eur. Respir. Rev.* **2023**, *32*, 220191. [CrossRef] [PubMed]
40. Bilezikian, J.P.; Binkley, N.; De Luca, H.F.; Fassio, A.; Formenti, A.M.; Fuleihan, G.E.-H.; Heijboer, A.C.; Giustina, A. Consensus and Controversial Aspects of Vitamin D and COVID-19. *J. Clin. Endocrinol. Metab.* **2023**, *108*, 1034–1042. [CrossRef] [PubMed]
41. Gibbons, J.B.; Norton, E.C.; McCullough, J.S.; Meltzer, D.O.; Lavigne, J.; Fiedler, V.C.; Gibbons, R.D. Association between Vitamin D Supplementation and COVID-19 Infection and Mortality. *Sci. Rep.* **2022**, *12*, 19397. [CrossRef]
42. Villasis-Keever, M.A.; López-Alarcón, M.G.; Miranda-Novales, G.; Zurita-Cruz, J.N.; Barrada-Vázquez, A.S.; González-Ibarra, J.; Martínez-Reyes, M.; Grajales-Muñiz, C.; Santacruz-Tinoco, C.E.; Martínez-Miguel, B.; et al. Efficacy and Safety of Vitamin D Supplementation to Prevent COVID-19 in Frontline Healthcare Workers. A Randomized Clinical Trial. *Arch. Med. Res.* **2022**, *53*, 423. [CrossRef]
43. Jolliffe, D.A.; Holt, H.; Greenig, M.; Talaei, M.; Perdek, N.; Pfeffer, P.; Vivaldi, G.; Maltby, S.; Symons, J.; Barlow, N.L.; et al. Effect of a Test-and-Treat Approach to Vitamin D Supplementation on Risk of All Cause Acute Respiratory Tract Infection and Covid-19: Phase 3 Randomised Controlled Trial (CORONAVIT). *BMJ* **2022**, *378*, e071230. [CrossRef]
44. Greiller, C.; Martineau, A. Modulation of the Immune Response to Respiratory Viruses by Vitamin D. *Nutrients* **2015**, *7*, 4240–4270. [CrossRef]
45. Keutmann, M.; Hermes, G.; Meinberger, D.; Roth, A.; Stemler, J.; Cornely, O.A.; Klatt, A.R.; Streichert, T. The Ratio of Serum LL-37 Levels to Blood Leucocyte Count Correlates with COVID-19 Severity. *Sci. Rep.* **2022**, *12*, 9447. [CrossRef] [PubMed]
46. Kudryashova, E.; Zani, A.; Vilmen, G.; Sharma, A.; Lu, W.; Yount, J.S.; Kudryashov, D.S. Inhibition of SARS-CoV-2 Infection by Human Defensin HNP1 and Retrocyclin RC-101. *J. Mol. Biol.* **2022**, *434*, 167225. [CrossRef] [PubMed]
47. Bilezikian, J.P.; Bikle, D.; Hewison, M.; Lazaretti-Castro, M.; Formenti, A.M.; Gupta, A.; Madhavan, M.V.; Nair, N.; Babalyan, V.; Hutchings, N.; et al. Mechanisms In Endocrinology: Vitamin D and COVID-19. *Eur. J. Endocrinol.* **2020**, *183*, R133. [CrossRef] [PubMed]
48. Michalakis, K.; Ilias, I. SARS-CoV -2 Infection and Obesity: Common Inflammatory and Metabolic Aspects. *Diabetes Metab. Syndr.* **2020**, *14*, 469. [CrossRef] [PubMed]
49. Gao, X.; Li, P.; Dai, S.; Wang, G.; Li, W.; Song, Z.; Zhu, L.; Zhu, S. Is Prior Bariatric Surgery Associated with Poor COVID-19 Outcomes? A Systematic Review and Meta-Analysis of Case-Control Studies. *J. Glob. Health* **2023**, *13*, 06012. [CrossRef] [PubMed]
50. Wood, G.C.; Benotti, P.N.; Fano, R.M.; Dove, J.T.; Rolston, D.D.; Petrick, A.T.; Still, C.D. Prior Metabolic Surgery Reduced COVID-19 Severity: Systematic Analysis from Year One of the COVID-19 Pandemic. *Heliyon* **2023**, *9*, e15824. [CrossRef] [PubMed]
51. Iannelli, A.; Bouam, S.; Schneck, A.-S.; Frey, S.; Zarca, K.; Gugenheim, J.; Alifano, M. The Impact of Previous History of Bariatric Surgery on Outcome of COVID-19. A Nationwide Medico-Administrative French Study. *Obes. Surg.* **2021**, *31*, 1455. [CrossRef] [PubMed]
52. Lewis, C.-A.; de Jersey, S.; Seymour, M.; Hopkins, G.; Hickman, I.; Osland, E. Iron, Vitamin B12, Folate and Copper Deficiency after Bariatric Surgery and the Impact on Anaemia: A Systematic Review. *Obes. Surg.* **2020**, *30*, 4542–4591. [CrossRef]
53. Bal, B.S.; Finelli, F.C.; Shope, T.R.; Koch, T.R. Nutritional Deficiencies after Bariatric Surgery. *Nat. Rev. Endocrinol.* **2012**, *8*, 544. [CrossRef]
54. Giusti, V.; Suter, M.; Héraïef, E.; Gaillard, R.C.; Burckhardt, P. Effects of Laparoscopic Gastric Banding on Body Composition, Metabolic Profile and Nutritional Status of Obese Women: 12-Months Follow-Up. *Obes. Surg.* **2004**, *14*, 239–245. [CrossRef]
55. Morales-Marroquin, E.; Xie, L.; Uppuluri, M.; Almandoz, J.P.; de la Cruz-Muñoz, N.; Messiah, S.E. Immunosuppression and Clostridioides (Clostridium) Difficile Infection Risk in Metabolic and Bariatric Surgery Patients. *J. Am. Coll. Surg.* **2021**, *233*, 223. [CrossRef] [PubMed]
56. Ebrahimzadeh, A.; Taghizadeh, M.; Milajerdi, A. Major Dietary Patterns in Relation to Disease Severity, Symptoms, and Inflammatory Markers in Patients Recovered from COVID-19. *Front. Nutr.* **2022**, *9*, 929384. [CrossRef] [PubMed]
57. Merino, J.; Joshi, A.D.; Nguyen, L.H.; Leeming, E.R.; Mazidi, M.; Drew, D.A.; Gibson, R.; Graham, M.S.; Lo, C.-H.; Capdevila, J.; et al. Diet Quality and Risk and Severity of COVID-19: A Prospective Cohort Study. *Gut* **2021**, *70*, 2096–2104. [CrossRef] [PubMed]
58. Kim, H.; Rebholz, C.M.; Hegde, S.; LaFiura, C.; Raghavan, M.; Lloyd, J.F.; Cheng, S.; Seidelmann, S.B.; Diets, P.-B. Pescatarian Diets and COVID-19 Severity: A Population-Based Case–Control Study in Six Countries. *BMJ Nutr. Prev. Health* **2021**, *4*, 257. [CrossRef]
59. Salazar-Robles, E.; Kalantar-Zadeh, K.; Badillo, H.; Calderón-Juárez, M.; García-Bárcenas, C.A.; Ledesma-Pérez, P.D.; Lerma, A.; Lerma, C. Association between Severity of COVID-19 Symptoms and Habitual Food Intake in Adult Outpatients. *BMJ Nutr. Prev. Health* **2021**, *4*, 469–478. [CrossRef]
60. Calder, P.C. Nutrition, Immunity and COVID-19. *BMJ Nutr. Prev. Health* **2020**, *3*, 74. [CrossRef]

61. Belanger, M.J.; Hill, M.A.; Angelidi, A.M.; Dalamaga, M.; Sowers, J.R.; Mantzoros, C.S. COVID-19 and Disparities in Nutrition and Obesity. *N. Engl. J. Med.* **2020**, *383*, e69. [CrossRef]
62. Yan, B.; Chu, H.; Yang, D.; Sze, K.-H.; Lai, P.-M.; Yuan, S.; Shuai, H.; Wang, Y.; Kao, R.Y.-T.; Chan, J.F.-W.; et al. Characterization of the Lipidomic Profile of Human Coronavirus-Infected Cells: Implications for Lipid Metabolism Remodeling upon Coronavirus Replication. *Viruses* **2019**, *11*, 73. [CrossRef]
63. Eiser, A.R. Could Dietary Factors Reduce COVID-19 Mortality Rates? Moderating the Inflammatory State. *J. Altern. Complement. Med.* **2021**, *27*, 176–178. [CrossRef]
64. Butler, M.J.; Barrientos, R.M. The Impact of Nutrition on COVID-19 Susceptibility and Long-Term Consequences. *Brain Behav. Immun.* **2020**, *87*, 53–54. [CrossRef]
65. Soltani, A.; Jaam, M.; Nazar, Z.; Stewart, D.; Shaito, A. Attitudes and Beliefs Regarding the Use of Herbs and Supplementary Medications with COVID-19: A Systematic Review. *Res. Social Adm. Pharm.* **2023**, *19*, 343–355. [CrossRef]
66. Hassan, S.S.U.; Abdel-Daim, M.M.; Behl, T.; Bungau, S. Natural Products for Chronic Diseases: A Ray of Hope. *Molecules* **2022**, *27*, 5573. [CrossRef]

nutrients

Review

The Impact of Chronic Stress Related to COVID-19 on Eating Behaviors and the Risk of Obesity in Children and Adolescents

Iwona Piątkowska-Chmiel [1], Paulina Krawiec [2], Karolina Joanna Ziętara [3], Piotr Pawłowski [3], Marzena Samardakiewicz [4], Elżbieta Pac-Kożuchowska [2] and Mariola Herbet [1,*]

1 Department of Toxicology, Faculty of Pharmacy, Medical University of Lublin, Jaczewskiego 8b Street, 20-090 Lublin, Poland; iwona.piatkowska-chmiel@umlub.pl
2 Department of Paediatrics and Gastroenterology, Medical University of Lublin, Al. Racławickie 1 Street, 20-059 Lublin, Poland; paulina.krawiec@umlub.pl (P.K.); elzbieta.pac-kozuchowska@umlub.pl (E.P.-K.)
3 Student Scientific Association at the Department of Psychology, Faculty of Medicine, Medical University of Lublin, 20-093 Lublin, Poland; kar.zietara@gmail.com (K.J.Z.); pawlowskipiotr56@gmail.com (P.P.)
4 Department of Psychology, Psychosocial Aspects of Medicine, Medical University of Lublin, Chodźki 7 Street, 20-093 Lublin, Poland; marzena.samardakiewicz@umlub.pl
* Correspondence: mariola.herbet@umlub.pl; Tel.: +48-81-448-74-04

Abstract: During the COVID-19 pandemic, an increase in the incidence of overweight and obesity in children was observed. It appears that unhealthy food choices, an unbalanced diet, and a sedentary lifestyle, as well as experiencing stress related to the pandemic, may be contributing to this disturbing trend. Chronic stress is a significant factor contributing to eating disorders and obesity in youngsters, involving medical, molecular, and psychological elements. Individuals under chronic stress often focus on appearance and weight, leading to negative body image and disrupted relationships with food, resulting in unhealthy eating behaviors. Chronic stress also impacts hormonal balance, reducing the satiety hormone leptin and elevating the appetite-stimulating hormone ghrelin, fostering increased hunger and uncontrolled snacking. Two systems, the hypothalamic–pituitary–adrenal axis and the sympathetic system with the adrenal medulla, are activated in response to stress, causing impaired secretion of noradrenaline and cortisol. Stress-related obesity mechanisms encompass oxidative stress, neuroinflammation, insulin resistance, and neurohormonal and neurotransmission disorders. Stress induces insulin resistance, elevating obesity risk by disrupting blood sugar regulation and fat storage. Stress also affects the gut microbiome, potentially influencing chronic inflammation and metabolic processes linked to obesity. In conclusion, chronic stress is a multifaceted risk factor for eating disorders and obesity in children, necessitating a comprehensive understanding of effective preventive and intervention strategies amid the escalating prevalence of childhood overweight and obesity.

Keywords: chronic stress; COVID-19 pandemic; overweight; obesity; children; adolescents; eating behaviors

1. The Impact of Chronic Stress Related to the COVID-19 Pandemic on Eating Behaviors and the Risk of Obesity in Childhood

The global consequences of the COVID-19 pandemic have profoundly affected the lives and health well-being of individuals worldwide. The extensive consequences of this unprecedented situation became visible in both the physical and psychological dimensions, giving rise to significant risks associated with non-communicable diseases. The implementation of social distancing measures during the pandemic has triggered a cascade of effects, including restrictions on outdoor physical activity, and alterations in eating habits and sleep patterns. These changes have proven to be pivotal factors contributing to the surge in cases of overweight and obesity, in adults and among children [1–4]. Growing evidence suggests that chronic stress associated with the COVID-19 pandemic has played a significant role in eating habits among the general population [1–7]. Stress is one of the important factors that can affect appetite food intake, eating patterns, and intestinal microbiome [8,9]. Stress-induced gut dysbiosis may influence chronic inflammation and metabolic processes that may be associated with obesity [10,11]. Furthermore, chronic stress may affect nourishment preferences toward the consumption of energy-dense foods, leading to the accumulation of abdominal adipose tissue and obesity [8]. A study involving Chinese students indicated that 24.9% of them experienced anxiety arising from the onset of COVID-19. This heightened anxiety was linked to factors such as quarantine, a lack of interpersonal contacts, the overwhelming influx of news, and exposure to false information—all contributing to an increased mental burden and a change in eating habits [12]. Clinical observations indicate that individuals dealing with chronic stress often develop a fixation on their physical appearance and weight, fostering a negative body image that disrupts their relationship with food. This discordance can give rise to maladaptive eating behaviors, such as emotional overeating—seeking solace in highly processed foods as a means of escaping stress or finding comfort. The consequences of such practices may manifest in the form of excessive calorie intake, ultimately leading to overweight and obesity. Regrettably, during the pandemic, there was a concerning rise in obesity, among children and adolescents [1]. The unintentional consequence of social distancing restrictions was a negative impact on physical activity opportunities [13,14]. A meta-analysis of 22 studies of 14,216 children aged 3–18 years indicated that during the COVID-19 pandemic, total engagement in physical activity in children was reduced by 20% (90% CI −34 to −4%) compared with the pre-pandemic period [13]. This negative phenomenon was most prominent in moderate-to-vigorous physical activity with a −28% change (90% CI: −41% to −13%) [14]. Isolation and an increase in sedentary lifestyles have played a key role in the development of excess weight among this demographic [15–18]. Moreover, it has been noticed that after the COVID-19 outbreak, screen time among children increased [19,20]. A recent report published by the CDC demonstrated that in the pediatric population in the United States between the pre-pandemic period and during the COVID-19 pandemic, the rate of body mass index almost doubled from 0.052 (95% CI = 0.051–0.052) to 0.100 (95% CI = 0.098–0.101) kg/m^2/month [1]. It should be also emphasized that the increase in the rate of BMI change was particularly apparent in those children and adolescents who experienced overweight and obesity before the pandemic [1,6].

A meta-analysis of twelve studies from eight countries also revealed a significant association between the lockdown during the COVID-19 pandemic and gain of body weight (MD 2.67; 95% CI 2.12–3.23; $p < 0.00001$) and increase in BMI (MD 0.77 kg/m^2; 95% CI 0.33–1.20; $p = 0.0006$) in school-age children and adolescents [2]. Based on the Children's Hospital of Philadelphia Care Network data, it has been reported that the most significant increase in the obesity rates during the COVID-19 pandemic was recorded in children aged 5 to 9 years and those who were Hispanic, non-Hispanic Black, publicly insured, or lower income [5]. A recent systematic review revealed that although in most analyzed studies there was no reported change in the number of meals per day among children during the pandemic, there was an increase in the amount of food consumed [21]. It seems that social isolation has contributed to a rise in the consumption of home-cooked meals, linked to an increased intake of legumes, vegetables, and fruit, along with reduced consumption of fast food [18,21]. However, a noticeable surge in snacking and the consumption of "comfort food" was observed during the pandemic [18,21]. The term "comfort food" refers to a variety of meals typically considered energy-dense, high-fat, and/or high-carbohydrate products believed to provide emotional comfort [22]. Consuming comfort food during the pandemic could be seen as a coping mechanism for dealing with emotional distress, negative emotions, insecurity, and loneliness [18,21]. Furthermore, the combination of social isolation and continuous media exposure, including television even during meals, has been linked to an elevated consumption of fried food, sweets, and sugar-sweetened beverages, accompanied by a decreased intake of fruits and vegetables [18]. There is a negative relationship between screen time and physical activity and an association between extra screen time and unhealthy eating habits [23].

As indicated, chronic stress is a complex risk factor in the context of eating disorders and obesity in children and adolescents. Understanding these multidimensional aspects is crucial in developing effective preventive and intervention strategies, as the increasing prevalence of overweight and obesity in children is one of the most important challenges for health care. This review provides a comprehensive analysis of the impact of chronic stress on obesity among children and adolescents, which has significantly increased during the COVID-19 pandemic. This study takes into account both psychological and biological pathomechanisms.

2. Psychological Aspects of Eating Disorders during the COVID-19 Pandemic

2.1. Mental Health, Pandemic Environment, and Eating Disorders

Stress is an element that forms the mental sphere of every human being, regardless of age, gender, stage of mental development, or social status. It has a chronic, relatively strong, and prolonged effect on the health and biopsychospiritual status of the individual, although it can also have an individually positive effect. It such a case, stress is called eustress, which generates a positive psychological response to the interacting stressors; this effect mainly depends on personal reasoning and definition of stress and its impact on well-being [24–27].

The perception of stress by the child and adolescent is related to many factors, various social exposures, the transition between phases of psychomotor development, changes in health, family, and school situations, imposed restrictions, and peer relationships. It should be noted that, according to recent studies, a child from the age of 2 is aware of the changes occurring in their environment. Such changes include the SARS-CoV-2 (COVID-19) virus pandemic and the associated reorganization of the environment, such as isolation, homeschooling, restrictions on peer relationships, health problems for children and adolescents themselves, as well as their loved ones and others, including the topics of dying and death [28–30].

During the SARS-CoV-2 pandemic, the occurrence of disorders related to food intake was observed in both children and adolescents. These included problems with appetite, choice of unhealthy foods, and changing dynamics in the development of previously diagnosed eating disorders. Some of the main elements associated with abnormal relationships with food are psychological and psychiatric aspects [31,32]. The pandemic period has led to increased levels of stress, anxiety, and depression in children and adolescents, which accompany the aforementioned problems [31–33].

Appetite disorders include a decrease, increase, or loss of the urge to take food. A study by Paiva et al. in Brazil found a statistically significant ($p < 0.001$) association between the presence of anxiety in children (related, among other things, to social isolation during a pandemic) and the occurrence of appetite problems. The presence of anxiety in children translated into a 3.12-fold increased risk of changes in appetite, which in the majority was manifested by increased food intake [34]. A study conducted in Spain by Lavigne-Cerván et al. found that more than half of the children and adolescents studied had moderate to high levels of anxiety, which coincided with the results of a study by Jiao et al. conducted in China that showed reduced or no appetite [35,36]. The National Child Traumatic Stress Network's guide to parents, published in 2020, highlighted the possibility of appetite-related disorders in children of all ages, indicating the nature of the problem [37]. Appetite is influenced, alongside increasing anxiety, by depressive symptoms, which are a challenge during pandemics [38,39]. Among children and adolescents suffering from depression, the majority show decreased or increased appetite, as well as weight disturbances [40–42].

Choosing the right food is the basis for a healthy diet. Children's and adolescents' dietary choices are influenced by parental eating habits, household habits, and the environment [43]. During the pandemic, access to smartphones, social isolation, and increasing stress led to the occurrence of dietary disorders. Increased intake of high-calorie foods, inter-meal snacking, and less frequent choice of healthy, low-calorie foods are caused by increasing stress and prolonged screen time [44–47].

According to the Diagnostic and Statistical Manual of Mental Disorders Fifth Edition, there are three specific eating disorders: anorexia nervosa (AN), bulimia nervosa (BN), and binge eating disorder (BED) [48]. The number of diagnoses of these disorders among children and adolescents increased during the COVID-19 pandemic, with AN being the predominant one. A cross-sectional study conducted in Canada by Agostino et al. confirmed a significant increase in new diagnoses of AN, rapid evaluation of disease markers, and hospitalizations for this condition [49–51]. Moreover, during the pandemic, there was also an increase in hospitalizations due to bulimia nervosa and other eating disorders [52].

It has to be highlighted that there is a mutual relationship between obesity and eating disorders, particularly bulimia nervosa and binge eating disorder [53]. Individuals with BED have a three to six times increased risk of obesity compared with the general population [54]. Although patients with BN may have normal weight, the lifetime prevalence of obesity in this group is about 33% [55]. In a cohort of Canadian adolescents, the prevalence of eating disorders was 9.3% in obese boys and 20.2% in obese girls compared with 2.1% and 8.4% of normal-weight boys and girls, respectively [56].

Obesity and eating disorders share a common pathophysiology that involves biological, environmental, behavioral, and cognitive determinants [53,57]. Possible shared genetic susceptibility for both disorders may be associated with at-mass and obesity-associated (FTO) gene polymorphisms [53]. Moreover, dysregulation of the hypothalamic–pituitary–adrenal axis (HPA) and gut dysbiosis play an important role in the pathophysiology of obesity and eating disorders [53]. Environmental risk factors identified for both disorders are, among others, weight teasing, internalization of unattainable beauty ideals portrayed on social media and television, social pressure and frequent criticism, bullying, and unhealthy family eating patterns [53,57]. Psychological determinants that may play a significant role in the development of obesity and eating disorders may include low self-esteem, negative self-evaluation, and high body dissatisfaction [53]. Body dissatisfaction, compromised interpersonal functioning, aberrant emotional regulation, and inappropriate weight control behaviors are also common factors contributing to both disorders [53,57].

Disordered eating symptomatology was exacerbatedduring the pandemic and were more pronounced in adolescents than in younger children. Individuals at increased risk for developing these conditions were also observed to exhibit their exponents. These conditions were also observed to be exhibited by individuals at an increased risk of developing them. Circumstances that may have a role in triggering eating disorders include increasing anxiety, social isolation, and more time spent on social media. This often resulted in disturbances in the perception of one's own body; excessive physical activity, fear of gaining weight, and inappropriate food intake were often observed as a result. These circumstances not only influenced the occurrence of eating disorder symptoms but also the motivation to recover [32,58–60]. Family members spending more time together at home as a result of social distancing and lockdown requirements reflected increased parental recognition of eating disorder symptoms in children, which may have had an impact on the increase in acute ED visits and hospital admissions. Some patients and their families linked the onset of lockdown as a trigger for hospital admission. The overall increase in hospital admissions for eating disorder exacerbations and prolonged stays has resulted in an increased demand for care related to the treatment of these disorders [61–63].

Table 1 provides information on the psychological and pedagogical interventions that are used in the treatment of eating disorders in children and adolescents.

Table 1. Psychological and pedagogical interventions used in eating disorder treatment based on [64–69].

Psychological and Pedagogical Interventions Used in Eating Disorder Treatment			
Anorexia Nervosa	**Bulimia Nervosa**	**Binge Eating Disorder**	**References**
• Family-based treatment (FBT)	• Family-based treatment (FBT)	• Family-based treatment (FBT)	
• Parent-focused treatment (PFT)	• Guided self-help cognitive behavioral therapy (gshCBT)	• Guided self-help cognitive behavioral therapy (gshCBT)	
• Systemic family therapy (SyFT)	• Individual supportive psychotherapy (SPT)	• Enhanced cognitive behavioral psychotherapy (CBT-E)	
• Enhanced cognitive behavioral psychotherapy (CBT-E)	• Cognitive behavioral therapy for adolescents (CBT-A)	• Dialectical behavior therapy (DBT)	
• Adolescent-focused therapy (AFT)	• Psychodynamic therapy (PDT)	• Interpersonal psychotherapy (IPT)	[64–69]
• Cognitive remediation therapy (CRT)	• Dialectical behavior therapy (DBT)	• Health education groups	
• Intensive multifamily therapy	• Interpersonal psychotherapy (IPT)		
• Create a multidisciplinary team including a pediatrician, psychiatrist, psychologist, family therapist, nurse, dietician, and social services. • Nutritional education for families. • Psychoeducation in nutrition. • Educational video materials.			

2.2. Prevention of Eating Disorders

In the face of the numerous described problems associated with stress-related eating disorders and obesity (covibesity) in a group of children and adolescents, preventive measures to counteract the psychosocial aspects that promote the development of these disorders become important.

The first group of recipients of the above-described measures should be the primary support system of the child or adolescent, namely the family environment. During the pandemic, adhering to the principles of social isolation, caregivers spent significantly more time at home, working remotely, supervising the learning of their offspring, and fulfilling the duties of daily life in the online system. As a result, adults became distanced from mundane pleasures and, consequently, a sense of disorganization, frustration, and anger was aroused, which was often unloaded on the child or adolescent [70–72]. Heightened emotional reactions from the parent, such as excessive criticism and hostility, can consequently lead to the child developing an eating disorder in the form of malnutrition or compulsive overeating [72]. Eating meals as a family can be a difficult psychological experience for a young person, and as a result, young people may restrict eating [71,72]. Available studies describe the relationship between the qualitative support of caregivers and the risk of eating disorders in their children [72]. In 2014, a program to support caregivers of people with eating disorders called Peer-Led Resilience (PiLAR) was introduced in Ireland. It was followed by an evaluation of its effects, which showed increased knowledge, skills, and improved psychological well-being of parents, positively influencing the quality of treatment and mental health of their children. Studies confirm the great importance of psychoeducation, self-help, and skills training in telepsychiatry systems in developing supportive attitudes of caregivers toward children with eating disorders and obesity [72,73]. For obesity, possible preventive actions for parents are shown in Figure 1.

Preventive actions aimed at children themselves are divided, according to the specialized literature, into two types: universal—carried out in all children regardless of the degree of assessed risk; selective—carried out in groups of high-risk children (e.g., dancers, athletes, and obese children) [74]. A meta-analysis and systematic review conducted by Chua, Tam, and Shorey confirms the effectiveness of implementing eating disorder and obesity prevention activities in school settings. These interventions are primarily aimed at reducing the internalization of beauty ideals, and thereby increasing students' self-esteem.

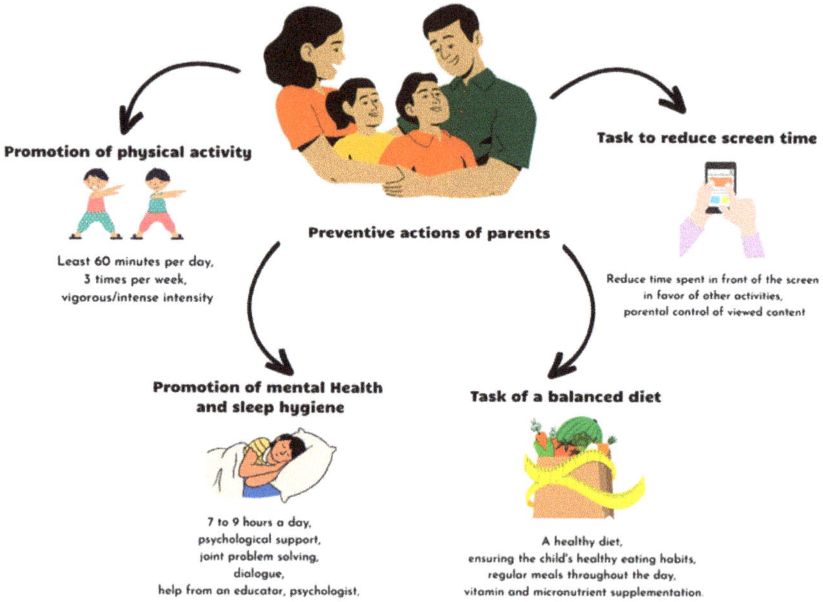

Figure 1. Possible covibesity prevention activities dedicated to parents [70–78].

These interventions are based on psychoeducation, training in the use of social media, and interpersonal training with an emphasis on the phenomena of teasing by peers. The study also indicates that girls are more prone to these activities than boys, due to their greater susceptibility to generating the cognitive illusion of a bad body image [75].

A major role in the prevention of eating disorders and obesity in the era of the COVID-19 pandemic is played by healthcare system workers, i.e., pediatricians, family physicians, and mental healthcare workers. Their roles and responsibilities are described in a narrative review by Singh et al. [76]. Pediatricians and family physicians conducting a periodic examination of a child or adolescent can recognize physical symptoms of stress and the patient's externalizing/internalizing emotional states. These professionals can then screen for mental disorders, including eating disorders and compulsive overeating, using brief, standardized screening tools such as the Children's Eating Disorders Inventory (EDI-C); the Eating Disorder Examination Questionnaire (EDE-Q); semi-structured interviews, such as the Children's Eating Disorder Examination (ChEDE); and online measures, such as the Development and Well-Being Assessment (DAWBA) [77,78]. The role of mental healthcare workers is to promote mental health (in the form of digital brochures, videos, etc., available online for caregivers, teachers, and children), increase mental health awareness and, most importantly, promote the practice of mental health hygiene [38,76].

3. Medical Aspects of Obesity in Children and Adolescents

3.1. Overview of Obesity in Childhood

According to the World Health Organization (WHO), overweight and obesity are disorders characterized by abnormal or excessive fat accumulation and associated with health hazards [79]. Table 2 presents diagnostic criteria of childhood obesity according to various definitions [80–85]. Abdominal or central obesity may be identified if a waist-to-height ratio is greater than 0.5 [15,86,87]. It has been also reported that the 90th percentile of waist circumference by age and sex for children and adolescents aged 6 to 18 years is the cutoff to identify central obesity in children and adolescents [88]. Obesity is one of the most serious challenges worldwide in all age groups [15,84,87]. Joint UNICEF/WHO/World Bank estimates demonstrated that globally, in 2020, the problem of overweight and obesity referred to 5.7% of children under 5 years of age, which aligns with 38.9 million children [89].

Table 2. Diagnostic criteria for childhood obesity [80–85].

Organization	Diagnostic Criteria for Obesity
World Health Organization [80,81]	Children under 5 years of age Overweight: age- and sex-adjusted weight-for-height >2 standard deviations above WHO Child Growth Standards median Obesity: age- and sex-adjusted weight-for-height >3 standard deviations above the WHO Child Growth Standards median Children between 5 and 19 years of age Overweight: age- and sex-adjusted BMI >1 standard deviation above the WHO Growth Reference median Obesity: age- and sex-adjusted BMI-for-age >2 standard deviations above the WHO Growth Reference median
U.S. Centers for Disease Control and Prevention (CDC) [82,83]	Children between 2 and 19 years of age Overweight: age- and sex-adjusted BMI ≥85th to <95th percentile on CDC Growth Charts Obesity: age- and sex-adjusted BMI ≥95th percentile on CDC Growth Charts Severe obesity: 120% of the 95th percentile of the age- and sex-adjusted BMI or greater, or BMI ≥35 kg/m^2 Children under 2 years of age CDC recommends the use of the WHO criteria
American Academy of Pediatrics (AAP) [84]	Severe obesity Class 2 obesity: age- and sex-adjusted BMI ≥ 35 to <40 kg/m^2 or ≥120 to 140% of the 95th percentile, whichever is lower Class 3 obesity: age- and sex-adjusted BMI ≥ 40 kg/m^2 or ≥140% of the 95th percentile, whichever is lower
International Obesity Task Force (IOTF) [85]	Overweight International age- and sex-specific BMI percentile cutoffs equivalent to BMI 25 kg/m^2 at age 18 years Obesity International age- and sex-specific BMI percentile cutoffs equivalent to BMI 30 kg/m^2 at age 18 years

The WHO European Regional Obesity Report 2022 indicated that in a population of children aged from 5 to 9 years, the prevalence of overweight was 29.5% and the prevalence of obesity was 11.6% [90]. Overweight occurred in 24.9% of older children and youth between 10 and 19 years of age. In this population, 7.1% of children were obese [90].

The National Health and Nutrition Examination Survey (NHANES) from 2017–2018 revealed that 16.1% of children and adolescents in the United States were overweight, and 19.3% had obesity, including approximately 30% with severe obesity [91].

Data from 2416 population-based studies in a pediatric population demonstrated that from 1975 to 2016 the increasing trend in mean BMI reached a plateau at high levels in many high-income countries, and a steep increase trend in east, south, and southeast Asia [92]. The highest prevalence of obesity exceeding 30% of the children's population was reported in some countries of the Pacific Ocean region [92].

According to the World Obesity Federation, the global prevalence of childhood obesity is predicted to rise from 10% to 20% in boys and from 8% to 18% in girls in the period 2020–2035 [16]. It is estimated that in 2035 there will be 383 million children living with obesity worldwide [16].

Obesity has a detrimental impact on children's health. Possible consequences of obesity in childhood are listed in Table 3 [15,84,87,93].

Table 3. Complications associated with obesity in childhood [15,84,87,93].

System	Disorder
Cardiovascular	Hypertension Left ventricle hypertrophy Premature atherosclerotic cardiovascular disease
Respiratory	Asthma Obstructive sleep apnea Sleep disorders Hypoventilation syndrome
Endocrine	Dyslipidemia Impaired glucose tolerance Type 2 diabetes Metabolic syndrome Polycystic ovary syndrome Impaired growth and puberty
Gastrointestinal	Metabolic dysfunction-associated steatotic liver disease Gastroesophageal reflux disease Cholelithiasis Constipation Micronutrient deficiencies
Musculoskeletal	Slipped capital femoral epiphysis Blount disease Fractures Pes planus
Skin	Acanthosis nigricans Striae Intertrigo Hidradenitis suppurativa Furunculosis
Renal	Enuresis Glomerulosclerosis
Dental	Dental caries Periodontal disease
Nervous system	Idiopathic intracranial hypertension
Psychosocial	Reduced self-esteem Depression Anxiety Disordered eating Internalizing disorders
Long-term complications	Adult obesity Coronary artery disease Type 2 diabetes Certain cancers Infertility Osteoarthritis

3.2. Neurochemical and Molecular Changes Induced by Chronic Stress

Although many studies demonstrate that chronic stress is associated with obesity [94–96], the exact mechanisms underlying this process have not yet been fully elucidated. Molecular and neurochemical changes associated with chronic stress can indeed exert a significant influence on the development and progression of obesity (Scheme 1). This influence arises from the intricate interplay between the stress response system and a range of physiological pathways linked to the regulation of appetite, metabolism, and energy storage.

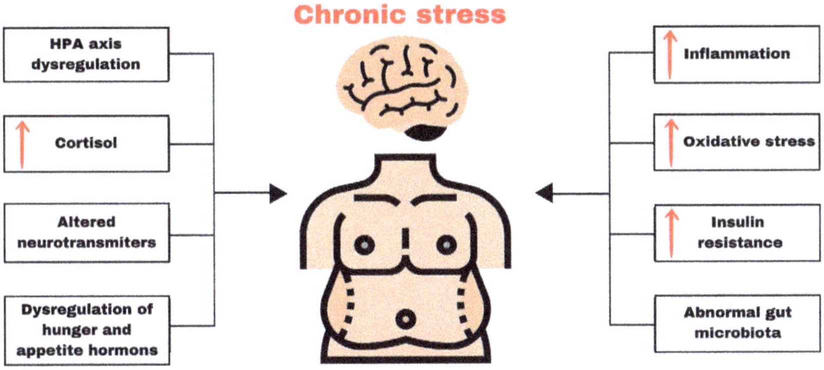

Scheme 1. Neurochemical and molecular alterations underlying chronic stress.

3.2.1. Chronic Activation of the HPA Axis

The stress system consists of the reactions of the sympathetic nervous system, which secretes noradrenaline and adrenaline, and the hypothalamic–pituitary–adrenal (HPA) axis, which releases cortisol, playing a major role in the physiology of stress reactions [97]. The effect of adrenal corticosteroids on the secretion of adrenocorticotropin is a complex phenomenon because it depends on the type of stress. As a result of acute stress, glucocorticosteroids (GCs) directly inhibit the activity of the hypothalamic–pituitary–adrenal axis (HPA), while as a result of exposure to chronic stress, the effect of these steroids on the brain is stimulating [98]. Activation of the HPA axis activates the secretion of cortisol, a steroid hormone that regulates behavior and food choices [96]. Chronic activation of the HPA axis during chronic stress may result in prolonged cortisol effects and a subsequent orexigenic response that may manifest itself in cravings for certain types of foods [8]. In addition, GCs increase the expression of corticotropin-releasing factor (CRF) in the central nucleus of the amygdala, the emotional processing center, enabling the recruitment of the chronic stress response network. Additionally, GCs increase the perception of pleasant or compulsive activities, which motivates the consumption of "comfort foods" [98]. In this way, GCs act systemically to lead to an increase in body fat, which, in turn, leads to energy storage, inhibition of catecholamines in the brainstem, and CRF expression in hypothalamic neurons. Studies have shown that an increase in the level of the stress hormone cortisol (glucocorticoid) plays a key role in the development of obesity. Glucocorticoid redistributes white adipose tissue in a given area and increases hunger attacks and the desire to eat high-energy food, consisting mainly of fat and sugars [99]. Additionally, it has been shown that high cortisol levels can also increase appetite [100]. In turn, people with abdominal obesity experience neuroendocrine disorders, which result in impaired functioning of the hypothalamus, pituitary, and adrenal glands. In vivo studies have shown that exposure to chronic stress reduces the weight gain of animals [101]. In turn, another study showed that when rats are stressed, they are more likely to eat lard and food rich in sugar [102]. However, it has been noticed that in humans, chronic stress may have a two-way effect—it may cause increased food consumption to ensure comfort of life, which leads to weight gain. It may also suppress appetite, reduce food intake, and weight loss. Studies have shown that depressed people who consume excessive amounts of food have reduced cerebrospinal CRF, catecholamine concentrations, and hypothalamic–pituitary–adrenal activity [98]. Disturbances in the activation of the hypothalamic–pituitary–adrenal axis are related to the development of obesity because the primary basis for the increased activation of this system is hyperinsulinemia, which causes functional hypoglycemia; this, in turn, leads to increased synthesis and release of ACTH, and then to excessive adrenal production of cortisol. Adam and Epel [100] related the reward-based stress eating model, which reveals the influence of cortisol on the consumption of high-calorie food and explains the importance of potential neuroendocrine mediators in the relationship between stress and eating. Both stress and eating stimulate the release of endogenous opioids, which, in turn, leads to a decrease in the activity of the HPA axis and weakens the stress response. Repeated stimulation of the reward pathway by stress stimuli induces stimulation of the HPA axis and reaching for food. This, in turn, may lead to neurobiological adaptation of the body and compulsive eating. Cortisol, in turn, may change the level of satisfaction achieved as a result of eating through neuroendocrine mediators such as insulin, leptin, and neuropeptide Y. Physiologically, the action of glucocorticoids is counteracted by insulin and leptin; however, in chronic stress this system becomes imbalanced. This may result in increased food intake, which may lead to the accumulation of fat tissue. Therefore, it seems that excessive activation of the HPA axis may be one of the pathomechanisms linking obesity and chronic stress. Chronic activation of the HPA axis can alter glucose metabolism, promote insulin resistance, and affect many appetite-related hormones and hypothalamic neuropeptides [103]. It has been shown that people exposed to chronic stress show a greater preference for and consumption of tasty, energy-rich foods rich in sugar and fat, which contributes to the development of obesity [104].

It has also been observed that the defensive reaction caused by stress leads to increased activation of the sympathetic nervous system and vasoconstriction in skeletal muscles [105]. It has been shown that increasing sympathetic activation may contribute to the development of obesity through stress factors. Therefore, it seems that this may be another pathophysiological link between chronic stress and obesity [105]. Because food is an inexpensive and readily available source of reward that provides short-term pleasure and relief from the discomfort associated with stress responses, negative reinforcement and stress may motivate eating to regulate stress responses.

Chronic glucocorticoid exposure increases the expression and activity of lipoprotein lipase within adipose tissue depots, facilitating fat storage, and this predominates within visceral adipose tissue since visceral depots express more GR than subcutaneous depots. The stromal vascular cells of visceral adipose express higher levels of 11β-hydroxysteroid dehydrogenase (11β-HSD-1), an enzyme that regenerates glucocorticoids from their inactive metabolites, further augmenting local glucocorticoid action in abdominal fat [106]. Moreover, cortisol can increase appetite and reduce the body's sensitivity to insulin, which leads to higher blood sugar levels [107]. Furthermore, cortisol plays a pivotal role in the activation of an enzyme called lipoprotein lipase (LPL), which results in the increased accumulation of triglycerides within fat cells, also known as adipocytes. It is essential to underscore that the level of cortisol is closely correlated with the extent of fat tissue within the body. What is of particular significance is the heightened sensitivity of intraabdominal fat tissue to the influence of cortisol, owing to the abundance of receptors for this hormone in comparison with other fat storage sites. Consequently, the adverse effects of excessive cortisol are notably pronounced in this region, thereby facilitating fat accumulation in the abdominal area [108].

Additionally, cortisol acts as an inhibitor of somatotropin, a growth hormone that typically carries out a lipolytic function under normal conditions, promoting the breakdown of fat. These intricate processes appear to be especially conspicuous in individuals afflicted by Cushing's syndrome, characterized by elevated levels of both cortisol and insulin, along with intra-abdominal obesity. Research conducted by Rebuffé-Scrive et al. in 1988 corroborates the association between hypercortisolemia, hyperinsulinemia, and intra-abdominal obesity [109]. Similarly, Rosmond et al. (1998) validated a substantial link between postprandial salivary cortisol levels and various health parameters and obesity-related indicators in men, such as BMI, waist-to-hip ratio (WHR), fasting glucose, insulin, triglycerides, cholesterol, and blood pressure [110].

3.2.2. Chronic Stress, Neurotransmitters, and Obesity

Potential mechanisms related to chronic stress and obesity include changes in the neurotransmitter system. Neurotransmitter disorder is a neurometabolic disorder that affects the way neurotransmitters are produced, broken down, or transported [111]. Neurotransmitter pathways include amino acids such as γ-aminobutyric acid (GABA), glutamate, and glycine, and monoamines such as adrenaline, noradrenaline, dopamine, and serotonin (5-HT). Chronic stress has been shown to affect brain motivation and habit-related areas [96,112]. Therefore, it is assumed that stress and reward circuits in the brain overlap and, by activating habit-based circuits, food cravings are increased by stress [96]. Catecholamine neurotransmitters such as dopamine (DA) and norepinephrine (NE) are found in brain neurons that are involved in the central nervous system's response to stress. It has been supported that stress is associated with an increase in the activity of neurons producing both DA and NE and leads to an increase in the synthesis of both transmitters [113]. It is known that catecholamines, which are secreted by the adrenal medulla and the sympathetic nervous system, exert a significant impact on the body's energy processes [114]. Lipolysis is initiated by catecholamines, which activate the process via β1- and β2-adrenergic receptors. This stimulation results in an elevation of intracellular cyclic adenosine monophosphate (cAMP) concentration due to cAMP-dependent activation of protein kinase [115]. HPA activation is associated with the activation of the mesolimbic dopaminergic system, a network strongly linked to the reward system. Dopamine has been linked to reward sensitivity, conditioning, and control of food abuse [103]. In people with obesity, the sensitivity of α- and β-adrenergic receptors in adipose tissue is modified. In this way, catecholamines influence the lipolytic process, leading to an increase in the storage of fat. The results of preclinical studies indicate an ambiguous response of DA to various stressful stimuli. It has been observed that acute, controlled physical stress causes increased DA outflow in the brain, while during long-term and uncontrolled exposure to the same stress factors, DA release is attenuated [116]. It is also known that chronic stress affects the neurotransmission of serotonin (5HT) in the brain. The neurotransmission of serotonin (5HT) in the brain is also known to be affected by chronic stress. Stress factors can activate raphe nuclei and rapidly increase 5HT release at efferent targets, such as the cortex, hypothalamus, and amygdala [117,118]. Therefore, chronic stress exposure may affect 5HT neurotransmission at postsynaptic targets, which may result in functional deficits in these brain regions, which, in turn, may result in abnormal behavioral patterns. Neurotransmitters such as serotonin and noradrenaline play an important role in the central nervous control of energy balance, which is why they are involved in obesity-related symptomatology. Many neurotransmitters have been shown to be involved in energy homeostasis by regulating food intake and/or energy expenditure; this phenomenon changes in the case of obesity [111]. However, whether changes in neurotransmitter metabolism are the cause or perhaps a consequence of overweight and obesity has not been fully clarified. This relationship is partially confirmed by the observation that weight gain may be caused by drugs targeting neurotransmission, such as atypical antipsychotics and antidepressants. This suggests that changes in neurotransmission may precede the onset of obesity. Another confirmation of the relationship may be the fact that the weight-loss properties of anti-obesity drugs are related to molecules acting as monoamine reuptake inhibitors or 5-HT receptor agonists and monoamine-releasing agents.

Chronic stressors have been shown to increase synaptic branching in the amygdala and anterior cingulate cortex while decreasing synaptic connections to the hippocampus and prefrontal regions [119]. Limbic regions, in turn, are involved in reward encoding and reward-based learning and feeding [120]. Repeated stressors that keep the stress system chronically activated can alter the brain's reward pathways related to food seeking, which, in turn, can lead to metabolic changes that promote fat storage. Body weight-dependent adaptations of neural pathways may therefore enhance food preferences, appetite, and food intake under conditions of chronic stress.

3.2.3. The Impact of Chronic Stress on the Neurohormonal Regulation of Appetite

The regulation of appetite is a highly complex process, finely tuned by the intricate interplay of various hormones and neurotransmitters in our body. These substances play a crucial role in governing our appetite, dictating when hunger strikes, when we reach a point of satisfaction, and even how our digestive system functions. Stress exerts a significant influence on our appetite and eating behaviors. In moments of acute stress, our appetite typically diminishes. However, the situation becomes markedly more intricate when we are under chronic stress. The prolonged periods of stress often trigger intense cravings and a pursuit of, as well as indulgence in, delicious, frequently high-calorie, and fatty foods. This phenomenon is closely intertwined with weight gain and an elevated risk of obesity.

The brain plays a central role in regulating appetite. The hypothalamus is a center that controls the feelings of hunger and satiety associated with eating. The arcuate nucleus produces various neurotransmitters such as neuropeptide Y (NPY) and agouti-related peptide (AgRP), which increase appetite, and alpha-melanocyte-stimulating hormone (α-MSH) and cocaine-and-amphetamine-regulated transcript (CART), which reduce appetite [121,122]. The paraventricular nucleus regulates the body's energy expenditure and controls food intake based on processing information received from the arcuate nucleus. Additionally, the paraventricular nucleus produces corticotropin-releasing hormone (CRH), which affects appetite [122].

Studies have shown that neuropeptide Y participates in the regulation of, and coping with, stress and maintains emotional homeostasis, preventing the behavioral consequences of stress and anxiety. Neuropeptide Y is also one of the main neuropeptides involved in increasing appetite, the action of which is inhibited by leptin [123].

It is also worth mentioning that leptin, a hormone produced by adipocytes, exerts a significant influence on the oxidation of fatty acids. Its concentration in the bloodstream is closely linked to the amount of body fat. Upon entering the bloodstream, leptin reaches the hypothalamus, where it exerts its effects on appetite regulation. Within the hypothalamus, leptin diminishes the production of neurotransmitters that promote appetite, such as NPY/AgRP, and enhances the production of neurotransmitters that suppress appetite, like POMC/CART [124]. This leads to a reduction in food consumption and an increase in the body's energy expenditure. Moreover, leptin not only affects the regulation of appetite and body weight but also plays an important role in maintaining the balance of blood glucose levels by increasing the sensitivity of peripheral tissues to insulin [125]. Stress is known to reduce leptin levels, a hormone released by adipose tissue that typically acts as a satiety signal to the brain. When leptin levels decrease, the body's hunger signals can intensify, leading to excess calorie consumption [126].

Bouillon-Minois et al. (2021) showed that leptin levels decrease in response to acute stress, mainly due to its effects on the HPA axis [127]. Experiments conducted in controlled laboratory environments indicate that leptin has the capacity to suppress the secretion of corticotropin-releasing hormone (CRH). This implies that the presence of leptin in the bloodstream might act as a regulator, potentially restraining the functioning of the HPA axis [128]. It should be noted that despite higher leptin levels in people experiencing increased stress, this does not always lead to the shutdown of the HPA axis (hypothalamic–pituitary–adrenal system). The brain appears to become less sensitive to the signals transmitted by leptin. This phenomenon makes it impossible to effectively inhibit the activity of the HPA axis and control appetite [129].

Another important hormone that plays a key role in regulating appetite and influences our eating habits is ghrelin. Ghrelin is known as the "hunger hormone" and is produced mainly by the stomach lining. Typically, after we have eaten, the levels of ghrelin in our bloodstream tend to decrease, which is a signal that helps promote a feeling of fullness or satiety. However, when there is a disruption in the normal regulation of ghrelin quality by a reduced suppression of ghrelin levels (in other words, ghrelin concentrations remain elevated) after a meal, it can play a role in the tendency to overeat. This disruption in ghrelin signaling may contribute to overconsumption of food [126,130]. Recent systematic studies have shown that ghrelin concentrations increase in response to both acute mental and physical stress [127]. Moreover, it has been proven that changes in the levels of the key stress hormone, cortisol, positively correlate with ghrelin levels both before and after meals in women struggling with obesity [131]. In situations of chronic stress or when stress reaches extremely high levels, the actions of cortisol and ghrelin may work together to contribute to stressful eating habits. This includes excessive consumption of very appetizing foods and eating despite not feeling hungry.

3.2.4. Chronic Stress, Inflammation, and Obesity

The relationship between chronic stress, inflammation, and obesity is complex. One of the key mechanisms through which chronic stress contributes to inflammation is the activation of the hypothalamic–pituitary–adrenal (HPA) axis. Prolonged exposure to elevated cortisol levels can disrupt the delicate balance of the immune system, resulting in an uneven distribution of pro-inflammatory and anti-inflammatory responses. This process involves immune cells, particularly macrophages, becoming active and producing pro-inflammatory cytokines like interleukin-6 (IL-6), interleukin-1β (IL-1β), and tumor necrosis factor α (TNF-α) [132]. Inflammatory processes play a disruptive role in the intricate mechanisms governing the metabolism of both fat and glucose, causing interference with the normal functioning of adipose tissue. Within the hypertrophic adipose tissue in obese individuals, nearly 50% of all cell types present are macrophages that exhibit an inflammatory phenotype. In the hypertrophic adipose tissue of obese individuals, almost half of all cell types are macrophages, and these macrophages display an inflammatory phenotype. The abundance of inflammatory macrophages within the adipose tissue significantly amplifies the ongoing state of inflammation, thereby exacerbating the intricate challenges linked to obesity. Consequently, this heightened inflammatory response often leads to a shift in fat accumulation toward visceral adipose tissue, ultimately culminating in the development of abdominal obesity, as highlighted in studies by Solinas et al. (2012) and Jager et al. (2007) [133,134]. A study by Lutz et al. (2016) showed that in people suffering from obesity, the content of macrophages in visceral adipose tissue was significantly higher compared with people with normal average body weight and amounted to as much as 12%. In people with normal average body weight, the macrophage content was only 4% [135]. Wannamethee et al. in 2007 confirmed that the concentration of IL-6 in portal veins correlates with the systemic inflammatory marker, C-reactive protein (CRP), in the blood of patients with visceral obesity [136]. Also, Illán-Gómez has confirmed a positive correlation between body mass index (BMI) and IL-6 levels in morbidly obese patients [137]. The results of a study by Arnardottir et al. (2012) confirm a strong correlation between IL-6 levels and BMI [138].

Regrettably, in the context of hypertrophied adipose tissue and the presence of pro-inflammatory mediators, the ability of adipocytes to produce not only adiponectin but also other beneficial adipokines is hampered. This, in turn, hinders the adipose tissue's capacity to fulfill its role in regulating anti-inflammatory, anti-hyperlipidemic, and insulin-sensitizing processes [139]. Adiponectin is a substance that works in several different ways to support metabolic health. One of its important actions is to inhibit the process of lipogenesis, i.e., the creation of new lipids, by blocking the action of the sterol-binding protein-1c. Moreover, adiponectin regulates the expression of pro-inflammatory genes by inhibiting the activity of the NF-κB transcription factor. Adiponectin also promotes the activity of two important transcription factors, PPAR-α(peroxisome proliferator-activated receptor alpha) and PPAR-γ (peroxisome proliferator-activated receptor gamma). These transcription factors are responsible for increasing the process of β-oxidation (fat burning in cells) and glucose transport into cells via the type 4 glucose transporter [140]. A study by Jaleel et al. (2006) showed that obese postmenopausal women have lower levels of adiponectin compared with women with normal body weight. Additionally, the study showed that obesity was associated with metabolic abnormalities, such as disturbances in blood lipid levels and changes in leptin levels [141].

3.2.5. Chronic Stress, Insulin Resistance, and Obesity

Chronic stress can negatively affect glucose homeostasis and lead to insulin resistance [142,143]. Research suggests that chronic stress may influence the development of insulin resistance in various ways. First, stress can increase the levels of hormones such as cortisol, which affect metabolism and may impair glucose regulation. Second, a notable link connection is observed between chronic stress and activation of the renin–angiotensin system (RAS) and insulin resistance. When this system is stimulated as a result of stress, there is a noticeable increase in the level of angiotensin II, a substance associated with it. Angiotensin II's detrimental impact extends to the inhibition of insulin-stimulated tyrosine phosphorylation (IST) via a mechanism dependent on mitogen-activated protein kinase (MAPK) [144]. Additionally, angiotensin II negatively affects the availability of nitric oxide (NO), subsequently impeding insulin signaling and diminishing insulin sensitivity [145]. Furthermore, angiotensin II can diminish the activity of PI3K, a crucial component of the insulin signaling pathway [144], and it disrupts the proper localization of Glut-4 on cell membranes [144].

Another important factor is excessive stimulation of the immune system as a result of chronic stress, which leads to an increase in the level of inflammatory mediators and pro-apoptotic factors. These phenomena have the potential to damage pancreatic cells. The consequence of this is disruption of the proper functioning of pancreatic cells, which, in turn, is associated with the risk of developing metabolic disorders, including insulin resistance [146]. In a 2014 study conducted by Parkulo, it was revealed that the structure and function of pancreatic islets in mice are detrimentally affected by long-term stress, which has detrimental effects on the structure and function of pancreatic islets in mice, resulting in the shrinkage of these structures and an increased risk of type 1 diabetes. Additionally, the research showed that chronic stress diminished the activity of genes responsible for the proliferation of pancreatic beta cells [147]. Furthermore, Huffman et al. (2013) found that individuals who experienced depression and persistent mental stress exhibited impaired beta cell function, directly impacting insulin availability and glucose homeostasis [148].

3.2.6. Chronic Stress, Oxidative Stress, and Obesity

In cases of chronic stress and inflammation, the equilibrium between the generation and removal of free oxygen radicals and their derivatives is disrupted, resulting in the onset of oxidative stress. Individuals enduring prolonged periods of stress often exhibit shifts in their eating habits. Stress can sometimes trigger "stress eating", leading people to reach for high-calorie and unhealthy snacks as a way to cope with their emotional burdens. This heightened intake of fats, carbohydrates, and unhealthy fatty acids can heighten oxidative stress via diverse biochemical pathways. These processes involve the overproduction of superoxide via NADPH oxidase activity, oxidative phosphorylation, spontaneous glyceraldehyde oxidation, activation of protein kinase C, and stimulation of polyol and hexosamine pathways [149]. These metabolic abnormalities lead to increased oxidative damage to cells, which may manifest as mitochondrial and DNA damage, as well as a decrease in adenosine triphosphate (ATP) levels and lipotoxicity [150]. As a result of increased oxidative damage, there is increased production of pro-inflammatory cytokines and acceleration of lipid peroxidation. This results in increased lipogenesis and altered insulin signaling, further enhancing the production of reactive oxygen species, and creating a detrimental feedback loop [151]. Consequently, oxidative stress is a significant contributor to the excessive accumulation of body energy substances such as glucose and fats in the liver, muscles, and adipose tissue, while stimulating mitochondrial and peroxisomal oxidation processes [152]. It is worth noting that people with obesity are more susceptible to oxidative damage due to depleted resources of antioxidants, such as superoxide dismutase (SOD), glutathione peroxidase (GPx), and catalase (CAT), as well as vitamins A, E, C, and β-carotene compared with people with normal body weight [153].

Reactive oxygen species (ROS) not only stimulate the secretion of pro-inflammatory cytokines, but also activate adhesion molecules and growth factors such as connective tissue growth factor, insulin-like growth factor-1 (IGF-I), platelet-derived growth factor, and vascular cell adhesion molecule-1. This process is possible owing to the action of transcription factors that respond to changes in the level of oxidation–reduction, especially by activating the NADPH (NOX) oxidase pathway and the NF-κB factor [154]. Moreover, ROS stimulates cytokine secretion through the apurin/apyrimidin endonuclease/redox factor-1 (APE/Ref-1-)-dependent pathway [155].

It is also important to recognize that adipose tissue serves as a source of bioactive molecules known as adipokines, which can play a significant role in oxidation–reduction balance. In people who are overweight or obese, changes in the level of adipokines are often observed, especially an increase in leptin concentration. Leptin, which is secreted mainly by adipocytes, plays a role in regulating appetite and the feeling of satiety. However, in people with obesity, leptin levels are often elevated, which may lead to resistance to its effects on appetite. At the same time, increased leptin can stimulate oxidative stress by increasing phagocyte activity, inducing the synthesis of pro-inflammatory cytokines such as TNF-α, IL-6, and IL-2, and increasing the levels of endothelial cell activation markers. This may contribute to the intensification of oxidative stress in the body of people with obesity [156].

In turn, adiponectin in people with obesity reaches low concentrations in blood serum, owing to which it loses its anti-inflammatory and anti-atherosclerotic properties [157]. At the same time, adiponectin is unable to effectively limit the release of reactive oxygen species (ROS) by low-density lipoproteins (LDLs), which results in increased oxidative stress in people struggling with overweight. In individuals with obesity, inflammatory adipocytokines like visfatin and resistin are released from adipose tissue. The concentrations of these adipocytokines exhibit a positive correlation with the quantity of adipose tissue and the occurrence of oxidative stress, as evidenced by studies conducted by Moschen et al. (2007) and Chen et al. (2010) [158,159].

3.2.7. Chronic Stress, Gut Dysbiosis, and Obesity

Stress is also a significant factor causing an imbalance in the gut microbiota homeostasis and gut dysbiosis [9]. The term "dysbiosis" refers to a disruption in microbiota equilibrium, leading to alterations in the composition, function, and activity of the microbial community [160]. Dysbiosis may be characterized as the loss of beneficial microorganisms, an increase in the abundance of pathobionts, and a reduction in microbial diversity [160,161]. Dysbiosis can be described as the depletion of beneficial microorganisms, a rise in the prevalence of pathogenic microorganisms, and a decrease in overall microbial diversity [160,161]. The gut microbiota plays a critical role in multiple physiological functions, including regulation of the HPA axis and sympathetic–adrenal medullary signaling, shaping the stress response [27]. Consequences of exposure to prolonged or chronic stressors in rodent models encompassed alterations in the microbiota composition and function, reduced stability of the microbiome, and increased microbial volatility [9]. Changes in the microbial community included in particular reduction in *Lactobacillaceae Akkermansia* and *Bifidobacteriacae* and an increase in pathobionts like *Clostridiacae, Escherichia*, and *Shigella* [9]. Gut dysbiosis appears to play a significant role in the development of obesity. As it has been described previously, stress is a significant factor affecting appetite and shaping dietary choices [10,11], while the diet is one of the most significant modifiable factors affecting gut microbiota homeostasis [162,163]. A high-fat diet is associated with increase in the *Firmicutes* to *Bacteriodetes* ratio, *Lactobacillus* spp., *Enterobacteriaceae, Bacteroidales, Bacteroides* spp., *Bifidobacterium* spp., and *Enterococcus* spp., and reduction in *Clostridia, Clostridium leptum*, and *Enterobacter* spp. in the gut microbiota composition [162]. Another implication of a high-fat diet is the low diversity of gut microbiota and the promotion of intestinal permeability [10,11]. It should be highlighted that the interaction between the diet and microbiota is bidirectional [10]. The gut microbiota with its metabolites may affect appetite regulation, reward system, or taste receptor expression [10]. Thus, it appears that stress, diet, and microbiota combine to form a virtuous cycle. Potential strategies for breaking this cycle may be focused on behavioral interventions aiming at diet modifications and stress reduction, as well as on therapeutic interventions leading to microbiota modulation.

The composition of gut microbiota in obese individuals may vary among different populations; however, in general, it is characterized by an elevated *Firmicutes* to *Bacteroidetes* ratio, and a decrease in *Bacteroides* spp., *Akkermansia* spp., and *Bifidobacterium* spp. [162,163]. Gut dysbiosis in the gut induced by a Western diet promotes obesogenic mechanisms, including the excessive harvesting extraction of energy from food and dysregulation of energy balance and gut hormones [160,163,164]. Moreover, alterations in gut microbiota may be associated with increased intestinal permeability, metabolic bacteriaemia, endotoxemia, and immune dysregulation propagating obesogenic inflammation [160,163,164]. However, further studies are needed to fully elucidate the complex role of the gut–brain axis in the pathogenesis of obesity.

4. Summary

The isolation policies necessary to limit the spread of SARS-CoV-2 infection have had an impact on the physical activity and eating habits of children and adolescents. Research has shown that the experience of chronic stress in the form of the COVID-19 pandemic plays an important role in many stages of the complex pathogenesis of obesity. Exposure to chronic stressors affects the neuroendocrine regulation of appetite and food preferences, and also promotes inflammation and intestinal dysbiosis, contributing to the development of obesity. It has also been shown that chronic stress influences the development of overweight and obesity through many mechanisms, such as the development of oxidative stress, neuroinflammation, insulin resistance, neurohormonal disorders, dysregulation of the HPA axis, and disorders in neurotransmission. Exposure to chronic stressors also affects the neuroendocrine regulation of appetite and food preferences and promotes inflammation and intestinal dysbiosis, contributing to the development of obesity. Obesity in children is a serious disease with serious consequences for mental, physical, and social health. It can lead to many diseases, including depression, coronary artery disease, type 2 diabetes, certain cancers, and osteoarthritis. Now more than ever, there is an urgent need to develop effective strategies to prevent overweight and obesity in the pediatric population and to develop effective multidisciplinary programs for the treatment of childhood obesity.

Author Contributions: Conceptualization, M.H. and I.P.-C.; methodology, M.H. and M.S.; investigation, K.J.Z. and P.P.; data curation, E.P.-K. and M.S.; writing—original draft preparation, P.K., I.P.-C., K.J.Z. and P.P.; writing—review and editing, M.H., P.K. and M.S.; visualization, K.J.Z. and P.P.; supervision, M.S., E.P.-K. and M.H.; and project administration, M.H. and I.P.-C. All authors have read and agreed to the published version of the manuscript.

Funding: This research received no external funding.

Institutional Review Board Statement: Not applicable.

Informed Consent Statement: Not applicable.

Data Availability Statement: Not applicable.

Conflicts of Interest: The authors declare no conflict of interest.

References

1. Lange, S.J.; Kompaniyets, L.; Freedman, D.S.; Kraus, E.M.; Porter, R.; Blanck, H.M.; Goodman, A.B. Longitudinal Trends in Body Mass Index before and during the COVID-19 Pandemic among Persons Aged 2–19 Years-United States, 2018–2020. *Morb. Mortal. Wkly. Rep.* **2021**, *70*, 1278–1283. [CrossRef] [PubMed]
2. Chang, T.H.; Chen, Y.C.; Chen, W.Y.; Chen, C.Y.; Hsu, W.Y.; Chou, Y.; Chang, Y.H. Weight Gain Associated with COVID-19 Lockdown in Children and Adolescents: A Systematic Review and Meta-Analysis. *Nutrients* **2021**, *13*, 3668. [CrossRef] [PubMed]

3. Anderson, L.N.; Yoshida-Montezuma, Y.; Dewart, N.; Jalil, E.; Khattar, J.; De Rubeis, V.; Carsley, S.; Griffith, L.E.; Mbuagbaw, L. Obesity and weight change during the COVID-19 pandemic in children and adults: A systematic review and meta-analysis. *Obes. Rev.* **2023**, *24*, e13550. [CrossRef] [PubMed]
4. Woolford, S.J.; Sidell, M.; Li, X.; Else, V.; Young, D.R.; Resnicow, K.; Koebnick, C. Changes in body mass index among children and adolescents during the COVID-19 pandemic. *JAMA* **2021**, *326*, 1434–1436. [CrossRef] [PubMed]
5. Jenssen, B.P.; Kelly, M.K.; Powell, M.; Bouchelle, Z.; Mayne, S.L.; Fiks, A.G. COVID-19 and changes in child obesity. *Pediatrics* **2021**, *147*, e2021050123. [CrossRef] [PubMed]
6. Vogel, M.; Geserick, M.; Gausche, R.; Beger, C.; Poulain, T.; Meigen, C.; Körner, A.; Keller, E.; Kiess, W.; Pfäffle, R. Age-and weight group-specific weight gain patterns in children and adolescents during the 15 years before and during the COVID-19 pandemic. *Int. J. Obes.* **2021**, *46*, 144–152. [CrossRef] [PubMed]
7. Jia, P.; Zhang, L.; Yu, W.; Yu, B.; Liu, M.; Zhang, D.; Yang, S. Impact of COVID-19 lockdown on activity patterns and weight status among youths in China: The COVID-19 Impact on Lifestyle Change Survey (COINLICS). *Int. J. Obes.* **2021**, *45*, 695–699. [CrossRef]
8. Torres, S.J.; Nowson, C.A. Relationship between stress, eating behavior, and obesity. *Nutrition* **2007**, *23*, 887–894. [CrossRef]
9. Leigh, S.J.; Uhlig, F.; Wilmes, L.; Sanchez-Diaz, P.; Gheorghe, C.E.; Goodson, M.S.; Kelley-Loughnane, N.; Hyland, N.P.; Cryan, J.F.; Clarke, G. The impact of acute and chronic stress on gastrointestinal physiology and function: A microbiota-gut-brain axis perspective. *J. Physiol.* **2023**, *601*, 4491–4538. [CrossRef]
10. Madison, A.; Kiecolt-Glaser, J.K. Stress, depression, diet, and the gut microbiota: Human-bacteria interactions at the core of psychoneuroimmunology and nutrition. *Curr. Opin. Behav. Sci.* **2019**, *28*, 105–110. [CrossRef]
11. Foster, J.A.; Rinaman, L.; Cryan, J.F. Stress & the gut-brain axis: Regulation by the microbiome. *Neurobiol. Stress.* **2017**, *7*, 124–136. [CrossRef] [PubMed]
12. Cao, W.; Fang, Z.; Hou, G.; Han, M.; Xu, X.; Dong, J.; Zheng, J. The psychological impact of the COVID-19 epidemic on college students in China. *Psychiatry Res.* **2020**, *287*, 112934. [CrossRef] [PubMed]
13. Neville, R.D.; Lakes, K.D.; Hopkins, W.G.; Tarantino, G.; Draper, C.E.; Beck, R.; Madigan, S. Global Changes in Child and Adolescent Physical Activity During the COVID-19 Pandemic: A Systematic Review and Meta-analysis. *JAMA Pediatr.* **2022**, *176*, 886–894. [CrossRef] [PubMed]
14. Warhadpande, M.; Sainz, K.; Jacobson, M.S. Effects of the COVID-19 Pandemic on Pediatric and Adolescent ASCVD Risk Factors. *Curr. Atheroscler. Rep.* **2023**, *25*, 591–596. [CrossRef] [PubMed]
15. Jebeile, H.; Kelly, A.S.; O'Malley, G.; Baur, L.A. Obesity in children and adolescents: Epidemiology, causes, assessment, and management. *Lancet Diabetes Endocrinol.* **2022**, *10*, 351–365. [CrossRef] [PubMed]
16. World Obesity Federation. World Obesity Atlas. 2023. Available online: https://data.worldobesity.org/publications/WOF-Obesity-Atlas-V5.pdf (accessed on 29 October 2023).
17. Zemrani, B.; Gehri, M.; Masserey, E.; Knob, C.; Pellaton, R. A hidden side of the COVID-19 pandemic in children: The double burden of undernutrition and overnutrition. *Int. J. Equity Health* **2021**, *20*, 44. [CrossRef] [PubMed]
18. Ruiz-Roso, M.B.; de Carvalho Padilha, P.; Mantilla-Escalante, D.C.; Ulloa, N.; Brun, P.; Acevedo-Correa, D.; Arantes Ferreira Peres, W.; Martorell, M.; Aires, M.T.; de Oliveira Cardoso, L.; et al. Covid-19 Confinement and Changes of Adolescent's Dietary Trends in Italy, Spain, Chile, Colombia and Brazil. *Nutrients* **2020**, *12*, 1807. [CrossRef]
19. Hedderson, M.M.; Bekelman, T.A.; Li, M.; Knapp, E.A.; Palmore, M.; Dong, Y.; Elliott, A.J.; Friedman, C.; Galarce, M.; Gilbert-Diamond, D.; et al. Environmental Influences on Child Health Outcomes Program. Trends in Screen Time Use among Children during the COVID-19 Pandemic, July 2019 through August 2021. *JAMA Netw Open* **2023**, *6*, e2256157. [CrossRef]
20. Ouyang, X.; Zhang, X.; Zhang, Q.; Gong, X.; Zhang, R. Preschool children's screen time during the COVID-19 pandemic: Associations with family characteristics and children's anxiety/withdrawal and approaches to learning. *Curr. Psychol.* **2023**, 1–15. [CrossRef]
21. Spence, C. Comfort food: A review. *Int. J. Gastron. Food Sci.* **2017**, *9*, 105–109. [CrossRef]
22. Pourghazi, F.; Eslami, M.; Ehsani, A.; Ejtahed, H.S.; Qorbani, M. Eating habits of children and adolescents during the COVID-19 era: A systematic review. *Front. Nutr.* **2022**, *9*, 1004953. [CrossRef] [PubMed]
23. Trott, M.; Driscoll, R.; Irlado, E.; Pardhan, S. Changes and correlates of screen time in adults and children during the COVID-19 pandemic: A systematic review and meta-analysis. *EClinicalMedicine* **2022**, *48*, 101452. [CrossRef] [PubMed]
24. Branson, V.; Dry, M.J.; Palmer, E.; Turnbull, D. The Adolescent Distress-Eustress Scale: Development and Validation. *Sage Open* **2019**, *9*, 2158244019865802. [CrossRef]
25. Branson, V.; Turnbull, D.; Dry, M.J.; Palmer, E. How Do Young People Experience Stress? A Qualitative Examination of the Indicators of Distress and Eustress in Adolescence. *Int. J. Stress. Manag.* **2019**, *26*, 321–329. [CrossRef]
26. Branson, V.; Palmer, E.; Dry, M.J.; Turnbull, D. A Holistic Understanding of the Effect of Stress on Adolescent Well-Being: A Conditional Process Analysis. *Stress. Health* **2019**, *35*, 626–641. [CrossRef] [PubMed]
27. Qian, L.; McWeeny, R.; Shinkaruk, C.; Baxter, A.; Cao, B.; Greenshaw, A.; Silverstone, P.; Pazderka, H.; Wei, Y. Child and Youth Mental Health and Wellbeing before and after Returning to In-Person Learning in Secondary Schools in the Context of COVID-19. *Front. Public. Health* **2023**, *11*, 1212297. [CrossRef] [PubMed]
28. Dalton, L.; Rapa, E.; Stein, A. Protecting the Psychological Health of Children through Effective Communication about COVID-19. *Lancet Child. Adolesc. Health* **2020**, *4*, 346–347. [CrossRef]

29. Liu, J.J.; Bao, Y.; Huang, X.; Shi, J.; Lu, L. Mental Health Considerations for Children Quarantined Because of COVID-19. *Lancet Child. Adolesc. Health* **2020**, *4*, 347–349. [CrossRef]
30. Imran, N.; Aamer, I.; Sharif, M.I.; Bodla, Z.H.; Naveed, S. Psychological Burden of Quarantine in Children and Adolescents: A Rapid Systematic Review and Proposed Solutions. *Pak. J. Med. Sci.* **2020**, *36*, 1106–1116. [CrossRef]
31. Magson, N.R.; Freeman, J.Y.A.; Rapee, R.M.; Richardson, C.E.; Oar, E.L.; Fardouly, J. Risk and Protective Factors for Prospective Changes in Adolescent Mental Health during the COVID-19 Pandemic. *J. Youth Adolesc.* **2021**, *50*, 44. [CrossRef]
32. Otto, A.K.; Jary, J.M.; Sturza, J.; Miller, C.A.; Prohaska, N.; Bravender, T.; Van Huysse, J. Medical Admissions among Adolescents with Eating Disorders during the COVID-19 Pandemic. *Pediatrics* **2021**, *148*, e2021052201. [CrossRef] [PubMed]
33. Devoe, D.J.; Han, A.; Anderson, A.; Katzman, D.K.; Patten, S.B.; Soumbasis, A.; Flanagan, J.; Paslakis, G.; Vyver, E.; Marcoux, G.; et al. The Impact of the COVID-19 Pandemic on Eating Disorders: A Systematic Review. *Int. J. Eat. Disord.* **2023**, *56*, 5–25. [CrossRef]
34. Paiva, E.D.; da Silva, L.R.; Machado, M.E.D.; de Aguiar, R.C.B.; da Silva Garcia, K.R.; Acioly, P.G.M. Child Behavior during the Social Distancing in the COVID-19 Pandemic. *Rev. Bras. Enferm.* **2021**, *74*, e20200762. [CrossRef] [PubMed]
35. Jiao, W.Y.; Wang, L.N.; Liu, J.; Fang, S.F.; Jiao, F.Y.; Pettoello-Mantovani, M.; Somekh, E. Behavioral and Emotional Disorders in Children during the COVID-19 Epidemic. *J. Pediatr.* **2020**, *221*, 264. [CrossRef] [PubMed]
36. Lavigne-Cerván, R.; Costa-López, B.; Juárez-Ruiz de Mier, R.; Real-Fernández, M.; Sánchez-Muñoz de León, M.; Navarro-Soria, I. Consequences of COVID-19 Confinement on Anxiety, Sleep and Executive Functions of Children and Adolescents in Spain. *Front. Psychol.* **2021**, *12*, 565516. [CrossRef] [PubMed]
37. Parent/Caregiver Guide to Helping Families Cope with the Coronavirus Disease 2019 | The National Child Traumatic Stress Network. Available online: https://www.nctsn.org/resources/parent-caregiver-guide-to-helping-families-cope-with-the-coronavirus-disease-2019 (accessed on 31 October 2023).
38. Chawla, N.; Tom, A.; Sen, M.S.; Sagar, R. Psychological Impact of COVID-19 on Children and Adolescents: A Systematic Review. *Indian J. Psychol. Med.* **2021**, *43*, 294–299. [CrossRef] [PubMed]
39. Courtney, D.; Watson, P.; Battaglia, M.; Mulsant, B.H.; Szatmari, P. COVID-19 Impacts on Child and Youth Anxiety and Depression: Challenges and Opportunities. *Can. J. Psychiatry* **2020**, *65*, 688–691. [CrossRef]
40. Grover, S.; Venkatesh Raju, V.; Sharma, A.; Shah, R.S. Depression in Children and Adolescents: A Review of Indian Studies. *Indian J. Psychol. Med.* **2019**, *41*, 216–227. [CrossRef]
41. Rice, F.; Riglin, L.; Lomax, T.; Souter, E.; Potter, R.; Smith, D.J.; Thapar, A.K.; Thapar, A. Adolescent and Adult Differences in Major Depression Symptom Profiles. *J. Affect. Disord.* **2019**, *243*, 175–181. [CrossRef]
42. Lindberg, L.; Hagman, E.; Danielsson, P.; Marcus, C.; Persson, M. Anxiety and Depression in Children and Adolescents with Obesity: A Nationwide Study in Sweden. *BMC Med.* **2020**, *18*, 1–9. [CrossRef]
43. Mahmood, L.; Flores-Barrantes, P.; Moreno, L.A.; Manios, Y.; Gonzalez-Gil, E.M. The Influence of Parental Dietary Behaviors and Practices on Children's Eating Habits. *Nutrients* **2021**, *13*, 1138. [CrossRef] [PubMed]
44. Moitra, P.; Madan, J. Impact of Screen Time during COVID-19 on Eating Habits, Physical Activity, Sleep, and Depression Symptoms: A Cross-Sectional Study in Indian Adolescents. *PLoS ONE* **2022**, *17*, e0264951. [CrossRef] [PubMed]
45. Kamaleddine, A.N.; Antar, H.A.; Ali, B.T.A.; Hammoudi, S.F.; Lee, J.; Lee, T.; Bhang, S.Y.; Chung, S.; Salameh, P. Effect of Screen Time on Physical and Mental Health and Eating Habits During COVID-19 Lockdown in Lebanon. *Psychiatry Investig.* **2022**, *19*, 220. [CrossRef] [PubMed]
46. Hill, D.C.; Moss, R.H.; Sykes-Muskett, B.; Conner, M.; O'Connor, D.B. Stress and Eating Behaviors in Children and Adolescents: Systematic Review and Meta-Analysis. *Appetite* **2018**, *123*, 14–22. [CrossRef] [PubMed]
47. Teixeira, M.T.; Vitorino, R.S.; da Silva, J.H.; Raposo, L.M.; Aquino, L.A.D.; Ribas, S.A. Eating Habits of Children and Adolescents during the COVID-19 Pandemic: The Impact of Social Isolation. *J. Human. Nutr. Diet.* **2021**, *34*, 670–678. [CrossRef] [PubMed]
48. Yu, Z.; Muehleman, V. Eating Disorders and Metabolic Diseases. *Int. J. Environ. Res. Public Health* **2023**, *20*, 2446. [CrossRef]
49. Agostino, H.; Burstein, B.; Moubayed, D.; Taddeo, D.; Grady, R.; Vyver, E.; Dimitropoulos, G.; Dominic, A.; Coelho, J.S. Trends in the Incidence of New-Onset Anorexia Nervosa and Atypical Anorexia Nervosa Among Youth During the COVID-19 Pandemic in Canada. *JAMA Netw. Open* **2021**, *4*, 2137395. [CrossRef]
50. Giacomini, G.; Elhadidy, H.S.M.A.; Paladini, G.; Onorati, R.; Sciurpa, E.; Gianino, M.M.; Borraccino, A. Eating Disorders in Hospitalized School-Aged Children and Adolescents during the COVID-19 Pandemic: A Cross-Sectional Study of Discharge Records in Developmental Ages in Italy. *Int. J. Environ. Res. Public Health* **2022**, *19*, 12988. [CrossRef]
51. Mizumoto, Y.; Sasaki, Y.; Sunakawa, H.; Tanese, S.; Shinohara, R.; Kurokouchi, T.; Sugimoto, K.; Seto, M.; Ishida, M.; Itagaki, K.; et al. Current Situation and Clinical Burden of Pediatricians for Children with Eating Disorders during the COVID-19 Pandemic. *Glob. Health Med.* **2023**, *5*, 122. [CrossRef]
52. Asch, D.A.; Buresh, J.; Allison, K.C.; Islam, N.; Sheils, N.E.; Doshi, J.A.; Werner, R.M. Trends in US Patients Receiving Care for Eating Disorders and Other Common Behavioral Health Conditions Before and During the COVID-19 Pandemic. *JAMA Netw. Open* **2021**, *4*, e2134913. [CrossRef]
53. Stabouli, S.; Erdine, S.; Suurorg, L.; Jankauskienė, A.; Lurbe, E. Obesity and Eating Disorders in Children and Adolescents: The Bidirectional Link. *Nutrients* **2021**, *13*, 4321. [CrossRef] [PubMed]
54. Breton, E.; Fotso Soh, J.; Booij, L. Immunoinflammatory processes: Overlapping mechanisms between obesity and eating disorders? *Neurosci. Biobehav. Rev.* **2022**, *138*, 104688. [CrossRef] [PubMed]

55. Villarejo, C.; Fernández-Aranda, F.; Jiménez-Murcia, S.; Peñas-Lledó, E.; Granero, R.; Penelo, E.; Tinahones, F.J.; Sancho, C.; Vilarrasa, N.; Montserrat-Gil de Bernabé, M.; et al. Lifetime obesity in patients with eating disorders: Increasing prevalence, clinical and personality correlates. *Eur. Eat. Disord. Rev.* **2012**, *20*, 250–254. [CrossRef] [PubMed]
56. Flament, M.F.; Henderson, K.; Buchholz, A.; Obeid, N.; Nguyen, H.N.; Birmingham, M.; Goldfield, G. Weight Status and DSM-5 Diagnoses of Eating Disorders in Adolescents From the Community. *J. Am. Acad. Child. Adolesc. Psychiatry* **2015**, *54*, 403–411.e2. [CrossRef] [PubMed]
57. Jebeile, H.; Lister, N.B.; Baur, L.A.; Garnett, S.P.; Paxton, S.J. Eating disorder risk in adolescents with obesity. *Obes. Rev.* **2021**, *22*, e13173. [CrossRef] [PubMed]
58. Vitagliano, J.A.; Jhe, G.; Milliren, C.E.; Lin, J.A.; Spigel, R.; Freizinger, M.; Woods, E.R.; Forman, S.F.; Richmond, T.K. COVID-19 and Eating Disorder and Mental Health Concerns in Patients with Eating Disorders. *J. Eat. Disord.* **2021**, *9*, 1–8. [CrossRef] [PubMed]
59. Spettigue, W.; Obeid, N.; Erbach, M.; Feder, S.; Finner, N.; Harrison, M.E.; Isserlin, L.; Robinson, A.; Norris, M.L. The Impact of COVID-19 on Adolescents with Eating Disorders: A Cohort Study. *J. Eat. Disord.* **2021**, *9*, 1–8. [CrossRef]
60. Graell, M.; Morón-Nozaleda, M.G.; Camarneiro, R.; Villaseñor, Á.; Yáñez, S.; Muñoz, R.; Martínez-Núñez, B.; Miguélez-Fernández, C.; Muñoz, M.; Faya, M. Children and Adolescents with Eating Disorders during COVID-19 Confinement: Difficulties and Future Challenges. *Eur. Eat. Disord. Rev.* **2020**, *28*, 864–870. [CrossRef]
61. Castellini, G.; Cassioli, E.; Rossi, E.; Innocenti, M.; Gironi, V.; Sanfilippo, G.; Felciai, F.; Monteleone, A.M.; Ricca, V. The Impact of COVID-19 Epidemic on Eating Disorders: A Longitudinal Observation of Pre versus Post Psychopathological Features in a Sample of Patients with Eating Disorders and a Group of Healthy Controls. *Int. J. Eat. Disord.* **2020**, *53*, 1855–1862. [CrossRef]
62. Shum, M.; Moreno, C.; Kamody, R.; McCollum, S.; Shabanova, V.; Loyal, J. The Evolving Needs of Children Hospitalized for Eating Disorders During the COVID-19 Pandemic. *Hosp. Pediatr.* **2022**, *12*, 696–702. [CrossRef]
63. Toulany, A.; Kurdyak, P.; Guttmann, A.; Stukel, T.A.; Fu, L.; Strauss, R.; Fiksenbaum, L.; Saunders, N.R. Acute Care Visits for Eating Disorders Among Children and Adolescents After the Onset of the COVID-19 Pandemic. *J. Adolesc. Health* **2022**, *70*, 42–47. [CrossRef] [PubMed]
64. Davis, L.E.; Attia, E. Recent Advances in Therapies for Eating Disorders. *F1000Research* **2019**, *8*. [CrossRef] [PubMed]
65. Bohon, C. Binge Eating Disorder in Children and Adolescents. *Child. Adolesc. Psychiatr. Clin. N. Am.* **2019**, *28*, 549. [CrossRef] [PubMed]
66. Hornberger, L.L.; Lane, M.A. Identification and Management of Eating Disorders in Children and Adolescents. *Pediatrics* **2021**, *147*, e2020040279. [CrossRef] [PubMed]
67. Mairs, R.; Nicholls, D. Assessment and Treatment of Eating Disorders in Children and Adolescents. *Arch. Dis. Child.* **2016**, *101*, 1168–1175. [CrossRef] [PubMed]
68. Herpertz-Dahlmann, B.; Dahmen, B. Children in Need—Diagnostics, Epidemiology, Treatment and Outcome of Early Onset Anorexia Nervosa. *Nutrients* **2019**, *11*, 1932. [CrossRef]
69. Gorrell, S.; Le Grange, D. Update on Treatments for Adolescent Bulimia Nervosa. *Child. Adolesc. Psychiatr. Clin. N. Am.* **2019**, *28*, 537. [CrossRef]
70. Brooks, S.K.; Webster, R.K.; Smith, L.E.; Woodland, L.; Wessely, S.; Greenberg, N.; Rubin, G.J. The Psychological Impact of Quarantine and How to Reduce It: Rapid Review of the Evidence. *Lancet* **2020**, *395*, 912–920. [CrossRef]
71. McCombie, C.; Austin, A.; Dalton, B.; Lawrence, V.; Schmidt, U. "Now It's Just Old Habits and Misery"-Understanding the Impact of the COVID-19 Pandemic on People With Current or Life-Time Eating Disorders: A Qualitative Study. *Front. Psychiatry* **2020**, *11*, 589225. [CrossRef]
72. Maunder, K.; McNicholas, F. Exploring Carer Burden amongst Those Caring for a Child or Adolescent with an Eating Disorder during COVID-19. *J. Eat. Disord.* **2021**, *9*, 1–8. [CrossRef]
73. PiLaR Program—An Evaluation. HSE National Clinical Program of Eating Disorders National Clinical Programme for Eating Disorders. 2019. Available online: https://www.bodywhys.ie/wp-content/uploads/2019/02/PilarReport_web.pdf (accessed on 31 October 2023).
74. Scapaticci, S.; Neri, C.R.; Marseglia, G.L.; Staiano, A.; Chiarelli, F.; Verduci, E. The Impact of the COVID-19 Pandemic on Lifestyle Behaviors in Children and Adolescents: An International Overview. *Ital. J. Pediatr.* **2022**, *48*, 1–17. [CrossRef] [PubMed]
75. Chua, J.Y.X.; Tam, W.; Shorey, S. Research Review: Effectiveness of Universal Eating Disorder Prevention Interventions in Improving Body Image among Children: A Systematic Review and Meta-Analysis. *J. Child. Psychol. Psychiatry* **2020**, *61*, 522–535. [CrossRef] [PubMed]
76. Singh, S.; Roy, D.; Sinha, K.; Parveen, S.; Sharma, G.; Joshi, G. Impact of COVID-19 and Lockdown on Mental Health of Children and Adolescents: A Narrative Review with Recommendations. *Psychiatry Res.* **2020**, *293*, 113429. [CrossRef] [PubMed]
77. Parker, M.N.; Lavender, J.M.; Schvey, N.A.; Tanofsky-Kraff, M. Practical Considerations for Using the Eating Disorder Examination Interview with Adolescents. *Adolesc. Health Med. Ther.* **2023**, *14*, 63–85. [CrossRef] [PubMed]
78. O'Logbon, J.; Newlove-Delgado, T.; McManus, S.; Mathews, F.; Hill, S.; Sadler, K.; Ford, T. How Does the Increase in Eating Difficulties According to the Development and Well-Being Assessment Screening Items Relate to the Population Prevalence of Eating Disorders? An Analysis of the 2017 Mental Health in Children and Young People Survey. *Int. J. Eat. Disord.* **2022**, *55*, 1777–1787. [CrossRef] [PubMed]

79. WHO World Health Organization. Obesity. Available online: https://www.who.int/health-topics/obesity#tab=tab_1 (accessed on 25 October 2023).
80. WHO World Health Organization. Obesity and Overweight. Available online: https://www.who.int/news-room/fact-sheets/detail/obesity-and-overweight (accessed on 25 October 2023).
81. de Onis, M.; Garza, C.; Onyango, A.W.; Martorell, R. WHO child growth standards. *Acta Paediatr.* **2006**, *95* (Suppl. S450), 3–101. [CrossRef]
82. Centers for Disease Control and Prevention. Defining Child BMI Categories. Available online: https://www.cdc.gov/obesity/basics/childhood-defining.html (accessed on 25 October 2023).
83. Grummer-Strawn, L.M.; Reinold, C.; Krebs, N.F. Centers for Disease Control and Prevention (CDC). Use of World Health Organization and CDC growth charts for children aged 0-59 months in the United States. *MMWR Recomm. Rep.* **2010**, *59*, 1–15.
84. Hampl, S.E.; Hassink, S.G.; Skinner, A.C.; Armstrong, S.C.; Barlow, S.E.; Bolling, C.F.; Avila Edwards, K.C.; Eneli, I.; Hamre, R.; Joseph, M.M.; et al. Clinical Practice Guideline for the Evaluation and Treatment of Children and Adolescents With Obesity. *Pediatrics* **2023**, *151*, e2022060640. [CrossRef]
85. Cole, T.J.; Lobstein, T. Extended international (IOTF) body mass index cut-offs for thinness, overweight and obesity. *Pediatr. Obes.* **2012**, *7*, 284–294. [CrossRef]
86. Eslami, M.; Pourghazi, F.; Khazdouz, M.; Tian, J.; Pourrostami, K.; Esmaeili-Abdar, Z.; Ejtahed, H.S.; Qorbani, M. Optimal cut-off value of waist circumference-to-height ratio to predict central obesity in children and adolescents: A systematic review and meta-analysis of diagnostic studies. *Front. Nutr.* **2023**, *9*, 985319. [CrossRef]
87. Lister, N.B.; Baur, L.A.; Felix, J.F.; Hill, A.J.; Marcus, C.; Reinehr, T.; Summerbell, C.; Wabitsch, M. Child and adolescent obesity. *Nat. Rev. Dis. Primers* **2023**, *18*, 24. [CrossRef] [PubMed]
88. Xi, B.; Zong, X.; Kelishadi, R.; Litwin, M.; Hong, Y.M.; Poh, B.K.; Steffen, L.M.; Galcheva, S.V.; Herter-Aeberli, I.; Nawarycz, T.; et al. International Waist Circumference Percentile Cutoffs for Central Obesity in Children and Adolescents Aged 6 to 18 Years. *J. Clin. Endocrinol. Metab.* **2020**, *105*, e1569–e1583. [CrossRef] [PubMed]
89. UNICEF/WHO/World Bank Joint Child Malnutrition Estimates: 2021 Edition Interactive Dashboard. New York (NY): United Nations Children's Fund. 2021. Available online: https://data.unicef.org/resources/joint-child-malnutrition-estimates-interactive-dashboard-2021/ (accessed on 29 October 2023).
90. World Health Organization (WHO). Regional Office for Europe. WHO European Regional Obesity Report. 2022. Available online: https://iris.who.int/bitstream/handle/10665/353747/9789289057738-eng.pdf?sequence=1&isAllowed=y (accessed on 29 October 2023).
91. Fryar, C.D.; Carroll, M.D.; Afful, J. Prevalence of Overweight, Obesity, and Severe Obesity among Children and Adolescents Aged 2–19 Years: United States, 1963–1965 through 2017–2018. NCHS Health E-Stats. 2020. Available online: https://www.cdc.gov/nchs/data/hestat/obesity-child-17-18/overweight-obesity-child-H.pdf (accessed on 29 October 2023).
92. NCD Risk Factor Collaboration (NCD-RisC). Worldwide trends in body-mass index, underweight, overweight, and obesity from 1975 to 2016: A pooled analysis of 2416 population-based measurement studies in 128 9 million children, adolescents, and adults. *Lancet* **2017**, *390*, 2627–2642. [CrossRef] [PubMed]
93. Skelton, J.A.; Klish, W.J. Overview of the Health Consequences of Obesity in Children and Adolescents. Available online: https://www.uptodate.com/contents/overview-of-the-health-consequences-of-obesity-in-children-and-adolescents?topicRef=5874&source=see_link#H1 (accessed on 29 October 2023).
94. Chen, Y.; Qian, L. Association between lifetime stress and obesity in Canadians. *Prev. Med.* **2012**, *55*, 464–467. [CrossRef] [PubMed]
95. Wardle, J.; Chida, Y.; Gibson, E.L.; Whitaker, K.L.; Steptoe, A. Stress and adiposity: A meta-analysis of longitudinal studies. *Obesity* **2011**, *19*, 771–778. [CrossRef] [PubMed]
96. Chao, A.; Grilo, C.M.; White, M.A.; Sinha, R. Food cravings mediate the relationship between chronic stress and body mass index. *J. Health Psychol.* **2015**, *20*, 721–729. [CrossRef] [PubMed]
97. Kumar, R.; Rizvi, M.R.; Saraswat, S. Obesity and Stress: A Contingent Paralysis. *Int. J. Prev. Med.* **2022**, *13*, 95.
98. Dallman, M.F.; Pecoraro, N.; Akana, S.F.; La Fleur, S.E.; Gomez, F.; Houshyar, H.; Bell, M.E.; Bhatnagar, S.; Laugero, K.D.; Manalo, S. Chronic stress and obesity: A new view of "comfort food". *Proc. Natl. Acad. Sci. USA* **2003**, *100*, 11696–11701. [CrossRef]
99. Fardet, L.; Fève, B. Systemic glucocorticoid therapy: A review of its metabolic and cardiovascular adverse events. *Drugs* **2014**, *74*, 1731–1745. [CrossRef]
100. Adam, T.C.; Epel, E.S. Stress, eating and the reward system. *Physiol. Behav.* **2007**, *91*, 449–458. [CrossRef]
101. Herbet, M.; Korga, A.; Gawrońska-Grzywacz, M.; Izdebska, M.; Piątkowska-Chmiel, I.; Poleszak, E.; Wróbel, A.; Matysiak, W.; Jodłowska-Jędrych, B.; Dudka, J. Chronic Variable Stress Is Responsible for Lipid and DNA Oxidative Disorders and Activation of Oxidative Stress Response Genes in the Brain of Rats. *Oxidative Med. Cell. Longev.* **2017**, *2017*, 7313090. [CrossRef] [PubMed]
102. Pecoraro, N.; Reyes, F.; Gomez, F.; Bhargava, A.; Dallman, M.F. Chronic stress promotes palatable feeding, which reduces signs of stress: Feedforward and feedback effects of chronic stress. *Endocrinology* **2004**, *145*, 3754–3762. [CrossRef] [PubMed]
103. Yau, Y.H.; Potenza, M.N. Stress and eating behaviors. *Minerva Endocrinol.* **2013**, *38*, 255–267. [PubMed]
104. Warne, J.P. Shaping the stress response: Interplay of palatable food choices, glucocorticoids, insulin and abdominal obesity. *Mol. Cell. Endocrinol.* **2009**, *300*, 137–146. [CrossRef] [PubMed]

105. Davy, K.P. The global epidemic obesity: Are we becoming more sympathetic? *Curr. Hypertens. Rep.* **2004**, *6*, 241–246. [CrossRef] [PubMed]
106. Nieuwenhuizen, A.G.; Rutters, F. The hypothalamic-pituitary-adrenal-axis in the regulation of energy balance. *Physiol. Behav.* **2008**, *94*, 169–177. [CrossRef] [PubMed]
107. Chao, A.M.; Jastreboff, A.M.; White, M.A.; Grilo, C.M.; Sinha, R. Stress, cortisol, and other appetite-related hormones: Prospective prediction of 6-month changes in food cravings and weight. *Obesity* **2017**, *25*, 713–720. [CrossRef]
108. Raber, J. Detrimental effects of chronic hypothalamic-pituitary-adrenal axis activation. From obesity to memory deficits. *Mol. Neurobiol.* **1998**, *18*, 1–22. [CrossRef]
109. Rebuffé-Scrive, M.; Krotkiewski, M.; Elfverson, J.; Björntorp, P. Muscle and adipose tissue morphology and metabolism in Cushing's syndrome. *J. Clin. Endocrinol. Metab.* **1988**, *67*, 1122–1128. [CrossRef]
110. Rosmond, R.; Dallman, M.F.; Björntorp, P. Stress-related cortisol secretion in men: Relationships with abdominal obesity and endocrine, metabolic and hemodynamic abnormalities. *J. Clin. Endocrinol. Metab.* **1998**, *83*, 1853–1859. [CrossRef]
111. Caron, A.; Jane Michael, N. New Horizons: Is Obesity a Disorder of Neurotransmission? *J. Clin. Endocrinol. Metab.* **2021**, *106*, e4872–e4886. [CrossRef] [PubMed]
112. Page, K.A.; Seo, D.; Belfort-DeAguiar, R.; Lacadie, C.; Dzuira, J.; Naik, S.; Amarnath, S.; Constable, R.T.; Sherwin, R.S.; Sinha, R. Circulating glucose levels modulate neural control of desire for high-calorie foods in humans. *J. Clin. Investig.* **2011**, *121*, 4161–4169. [CrossRef] [PubMed]
113. Institute of Medicine (US) Committee on Military Nutrition Research; Marriott, B.M. (Eds.) *Food Components to Enhance Performance: An Evaluation of Potential Performance-Enhancing Food Components for Operational Rations*; National Academies Press (US): Cambridge, MA, USA, 1994.
114. Zouhal, H.; Jacob, C.; Delamarche, P.; Gratas-Delamarche, A. Catecholamines and the effects of exercise, training and gender. *Sports Med.* **2008**, *38*, 401–423. [CrossRef] [PubMed]
115. Zouhal, H.; Lemoine-Morel, S.; Mathieu, M.E.; Casazza, G.A.; Jabbour, G. Catecholamines and obesity: Effects of exercise and training. *Sports Med. (Auckl. N.Z.)* **2013**, *43*, 591–600. [CrossRef] [PubMed]
116. Kumar, A.; Rinwa, P.; Kaur, G.; Machawal, L. Stress: Neurobiology, consequences and management. *J. Pharm. Bioallied Sci.* **2013**, *5*, 91–97. [CrossRef] [PubMed]
117. Hale, M.W.; Lowry, C.A. Functional topography of midbrain and pontine serotonergic systems: Implications for synaptic regulation of serotonergic circuits. *Psychopharmacology* **2011**, *213*, 243–264. [CrossRef]
118. Natarajan, R.; Northrop, N.A.; Yamamoto, B.K. Protracted effects of chronic stress on serotonin-dependent thermoregulation. *Stress (Amst. Neth.)* **2015**, *18*, 668–676. [CrossRef]
119. Berthoud, H.R. The neurobiology of food intake in an obesogenic environment. *Proc. Nutr. Soc.* **2012**, *71*, 478–487. [CrossRef]
120. Sinha, R. Chronic stress, drug use, and vulnerability to addiction. *Ann. N. Y. Acad. Sci.* **2008**, *1141*, 105–130. [CrossRef]
121. Bloom, S. Hormonal regulation of appetite. *Obes. Rev.* **2007**, *8*, 63–65. [CrossRef]
122. Takeda, E.; Terao, J.; Nakaya, Y.; Miyamoto, K.; Baba, Y.; Chuman, H.; Kaji, R.; Ohmori, T.; Rokutan, K. Stress control and human nutrition. *J. Med. Investig.* **2004**, *51*, 139–145. [CrossRef] [PubMed]
123. Zahorska-Markiewicz, B.; Obuchowicz, E.; Waluga, M.; Tkacz, E.; Herman, Z.S. Neuropeptide Y in obese women during treatment with adrenergic modulation drugs. *Med. Sci. Monit.* **2001**, *7*, 403–408. [PubMed]
124. Baver, S.B.; Hope, K.; Guyot, S.; Bjørbaek, C.; Kaczorowski, C.; O'Connell, K.M. Leptin modulates the intrinsic excitability of AgRP/NPY neurons in the arcuate nucleus of the hypothalamus. *J. Neurosci. Off. J. Soc. Neurosci.* **2014**, *34*, 5486–5496. [CrossRef] [PubMed]
125. Park, H.-K.; Ahima, R.S. Physiology of leptin: Energy homeostasis, neuroendocrine function and metabolism. *Metabolism* **2015**, *64*, 24–34. [CrossRef] [PubMed]
126. Klok, M.; Jakobsdottir, S.; Drent, M. The role of leptin and ghrelin in the regulation of food intake and body weight in humans: A review. *Obes. Rev.* **2007**, *8*, 21–34. [CrossRef] [PubMed]
127. Bouillon-Minois, J.B.; Trousselard, M.; Thivel, D.; Benson, A.C.; Schmidt, J.; Moustafa, F.; Bouvier, D.; Dutheil, F. Leptin as a Biomarker of Stress: A Systematic Review and Meta-Analysis. *Nutrients* **2021**, *13*, 3350. [CrossRef] [PubMed]
128. Heiman, M.L.; Ahima, R.S.; Craft, L.S.; Schoner, B.; Stephens, T.W.; Flier, J.S. Leptin inhibition of the hypothalamic-pituitaryadrenal axis in response to stress. *Endocrinology* **1997**, *138*, 3859–3863. [CrossRef] [PubMed]
129. Benomar, Y.; Wetzler, S.; Larue-Achagiotis, C.; Djiane, J.; Tome, D.; Taouis, M. In vivo leptin infusion impairs insulin and leptin signalling in liver and hypothalamus. *Mol. Cell Endocrinol.* **2005**, *242*, 59–66. [CrossRef]
130. le Roux, C.W.; Patterson, M.; Vincent, R.P.; Hunt, C.; Ghatei, M.A.; Bloom, S.R. Postprandial plasma ghrelin is suppressed proportional to meal calorie content in normal-weight but not obese subjects. *J. Clin. Endocrinol. Metab.* **2005**, *90*, 1068–1071. [CrossRef]
131. Sarker, M.R.; Franks, S.; Caffrey, J. Direction of post-prandial ghrelin response associated with cortisol response, perceived stress and anxiety, and self-reported coping and hunger in obese women. *Behav. Brain Res.* **2013**, *257*, 197–200. [CrossRef]
132. Osborn, O.; Olefsky, J.M. The cellular and signaling networks linking the immune system and metabolism in disease. *Nat. Med.* **2012**, *18*, 363–374. [CrossRef] [PubMed]
133. Solinas, G.; Karin, M. JNK1 and IKKbeta: Molecular links between obesity and metabolic dysfunction. *FASEB J. Off. Publ. Fed. Am. Soc. Exp. Biol.* **2010**, *24*, 2596–2611. [CrossRef]

134. Jager, J.; Grémeaux, T.; Cormont, M.; Le Marchand-Brustel, Y.; Tanti, J.-F. Interleukin-1beta-induced insulin resistance in adipocytes through down-regulation of insulin receptor substrate-1 expression. *Endocrinology* **2007**, *148*, 241–251. [CrossRef] [PubMed]
135. Lutz, S.Z.; Peter, A.; Machicao, F.; Lamprinou, A.; Machann, J.; Schick, F.; Königsrainer, I.; Königsrainer, A.; Fritsche, A.; Staiger, H.; et al. Genetic Variation in the 11β-hydroxysteroid-dehydrogenase 1 Gene Determines NAFLD and Visceral Obesity. *J. Clin. Endocrinol. Metab.* **2016**, *101*, 4743–4751. [CrossRef] [PubMed]
136. Wannamethee, S.G.; Whincup, P.H.; Rumley, A.; Lowe, G.D. Inter-relationships of interleukin-6, cardiovascular risk factors and the metabolic syndrome among older men. *J. Thromb. Haemost.* **2007**, *5*, 1637–1643. [CrossRef] [PubMed]
137. Illán-Gómez, F.; Gonzálvez-Ortega, M.; Orea-Soler, I.; Alcaraz-Tafalla, M.S.; Aragón-Alonso, A.; Pascual-Díaz, M.; Pérez-Paredes, M.; Lozano-Almela, M.L. Obesity and inflammation: Change in adiponectin, C-reactive protein, tumour necrosis factor-alpha and interleukin-6 after bariatric surgery. *Obes. Surg.* **2012**, *22*, 950–955. [CrossRef] [PubMed]
138. Arnardottir, E.S.; Maislin, G.; Schwab, R.J.; Staley, B.; Benediktsdottir, B.; Olafsson, I.; Juliusson, S.; Romer, M.; Gislason, T.; Pack, A.I. The interaction of obstructive sleep apnea and obesity on the inflammatory markers C-reactive protein and interleukin-6: The Icelandic Sleep Apnea Cohort. *Sleep* **2012**, *35*, 921–932. [CrossRef]
139. Guilherme, A.; Virbasius, J.V.; Puri, V.; Czech, M.P. Adipocyte dysfunctions linking obesity to insulin resistance and type 2 diabetes. *Nat. Rev. Mol. Cell Biol.* **2008**, *9*, 367–377. [CrossRef]
140. Kadowaki, T.; Yamauchi, T.; Kubota, N.; Hara, K.; Ueki, K.; Tobe, K. Adiponectin and adiponectin receptors in insulin resistance, diabetes, and the metabolic syndrome. *J. Clin. Investig.* **2006**, *116*, 1784–1792. [CrossRef]
141. Jaleel, F.; Jaleel, A.; Rahman, M.; Alam, E. Comparison of adiponectin, leptin and blood lipid levels in normal and obese postmenopause women. *J. Pak. Med. Assoc.* **2006**, *56*, 391–394.
142. Yaribeygi, H.; Sathyapalan, T.; Atkin, S.L.; Sahebkar, A. Molecular mechanisms linking oxidative stress and Diabetes mellitus. *Oxid. Med. Cell Longev.* **2020**, *2020*, 8609213. [CrossRef]
143. Saad, M.; Santos, A.; Prada, P. Linking gut microbiota and inflammation to obesity and insulin resistance. *Physiology* **2016**, *31*, 283–293. [CrossRef] [PubMed]
144. Zhou, M.S.; Schulman, I.H.; Zeng, Q. Związek między układem renina-angiotensyna a insulinoopornością: Implikacje dla chorób sercowo-naczyniowych. *Vasc. Med.* **2012**, *17*, 330–341. [CrossRef] [PubMed]
145. Steinberg, H.O.; Brechtel, G.; Johnson, A.; Fineberg, N.; Baron, A.D. Insulin-mediated skeletal muscle vasodilation is nitric oxide dependent. A novel action of insulin to increase nitric oxide release. *J. Clin. Investig.* **1994**, *94*, 1172–1179. [CrossRef] [PubMed]
146. Hackett, R.A.; Steptoe, A. Type 2 diabetes mellitus and psychological stress—A modifiable risk factor. *Nat. Rev. Endocrinol.* **2017**, *13*, 547. [CrossRef] [PubMed]
147. Parkulo, T. The Effects of Chronic Stress and Exercise on Mouse Pancreatic Islet of Langerhans Morphology and Muscle Atrophy Gene Expression. Graduate Thesis, West Virginia University, Morgantown, WV, USA, 2014.
148. Huffman, F.G.; Vallasciani, M.; Vaccaro, J.A.; Exebio, J.C.; Zarini, G.G.; Nayer, A.; Ajabshir, S. The association of depression and perceived stress with beta cell function between African and Haitian Americans with and without type 2 diabetes. *J. Diabetes Mellit.* **2013**, *3*, 236–243. [CrossRef] [PubMed]
149. Dandona, P.; Ghanim, H.; Chaudhuri, A.; Dhindsa, S.; Kim, S.S. Macronutrient intake induces oxidative and inflammatory stress: Potential relevance to atherosclerosis and insulin resistance. *Exp. Mol. Med.* **2010**, *42*, 245. [CrossRef]
150. Čolak, E. New markers of oxidative damage to macromolecules. *J. Med. Biochem.* **2008**, *27*, 1. [CrossRef]
151. Martyn, J.A.; Kaneki, M.; Yasuhara, S. Obesity-induced insulin resistance and hyperglycemia: Etiologic factors and molecular mechanisms. *Anesthesiology* **2008**, *109*, 137–148. [CrossRef]
152. Marseglia, L.; Manti, S.; D'Angelo, G.; Nicotera, A.; Parisi, E.; Di Rosa, G.; Gitto, E.; Arrigo, T. Oxidative stress in obesity: A critical component in human diseases. *Int. J. Mol. Sci.* **2014**, *16*, 378–400. [CrossRef]
153. Amirkhizi, F.; Siassi, F.; Minaie, S.; Djalali, M.; Rahimi, A.; Chamari, M. Is obesity associated with increased plasma lipid peroxidación and oxidative stress in women. *ARYA Atheroscler. J.* **2007**, *2*, 189–192.
154. Shoelson, S.E.; Herrero, L.; Naaz, A. Obesity, inflammation, and insulin resistance. *Gastroenterology* **2007**, *132*, 2169–2180. [CrossRef] [PubMed]
155. Frossi, B.; de Carli, M.; Daniel, K.C.; Rivera, J.; Pucillo, C. Oxidative stress stimulates IL-4 and IL-6 production in mast cells by an APE/Ref-1-dependent pathway. *Eur. J. Immunol.* **2003**, *33*, 2168–2177. [CrossRef] [PubMed]
156. Singh, P.; Sharma, P.; Sahakyan, K.R.; Davison, D.E.; Sert-Kuniyoshi, F.H.; Romero-Corral, A.; Swain, J.M.; Jensen, M.D.; Lopez-Jimenez, F.; Kara, T.; et al. Differential effects of leptin on adiponectin expression with weight gain versus obesity. *Int. J. Obes.* **2016**, *40*, 266–274. [CrossRef] [PubMed]
157. Iwan-Zietek, I.; Ruszkowska-Ciastek, B.; Michalska, M.; Overskaug, E.; Goralczyk, K.; Dabrowiecki, S.; Rosc, D. Association of adiponectin and leptin-to-adiponectin ratio with the function of platelets in morbidly obese patients. *J. Physiol. Pharmacol. Off. J. Pol. Physiol. Soc.* **2016**, *67c*, 555–561.
158. Moschen, A.R.; Kaser, A.; Enrich, B.; Mosheimer, B.; Theurl, M.; Niederegger, H.; Tilg, H. Visfatin an adipocytokine with proinflammatory and immunomodulating properties. *J. Immunol.* **2007**, *178*, 1748–1758. [CrossRef] [PubMed]
159. Chen, C.; Jiang, J.; Lü, J.M.; Chai, H.; Wang, X.; Lin, P.H.; Yao, Q. Resistin decreases expression of endothelial nitric oxide synthase through oxidative stress in human coronary artery endothelial cells. *Am. J. Physiol. HeartCirc. Physiol.* **2010**, *299*, 193–201. [CrossRef] [PubMed]

160. DeGruttola, A.K.; Low, D.; Mizoguchi, A.; Mizoguchi, E. Current Understanding of Dysbiosis in Disease in Human and Animal Models. *Inflamm. Bowel Dis.* **2016**, *22*, 1137–1150. [CrossRef]
161. Petersen, C.; Round, J.L. Defining dysbiosis and its influence on host immunity and disease. *Cell Microbiol.* **2014**, *16*, 1024–1033. [CrossRef]
162. Geng, J.; Ni, Q.; Sun, W.; Li, L.; Feng, X. The links between gut microbiota and obesity and obesity related diseases. *Biomed. Pharmacother.* **2022**, *147*, 112678. [CrossRef]
163. Amabebe, E.; Robert, F.O.; Agbalalah, T.; Orubu, E.S.F. Microbial dysbiosis-induced obesity: Role of gut microbiota in homoeostasis of energy metabolism. *Br. J. Nutr.* **2020**, *123*, 1127–1137. [CrossRef] [PubMed]
164. Valdes, A.M.; Walter, J.; Segal, E.; Spector, T.D. Role of the gut microbiota in nutrition and health. *BMJ* **2018**, *361*, k2179. [CrossRef] [PubMed]

nutrients

MDPI

Article

How COVID-19 Pandemic Restrictions Affected Kuwaiti College Students' Anthropometry, Lifestyle Behaviors, and Dietary Habits

Ahmad R. Al-Haifi [1], Nayef Y. Bumaryoum [1], Balqees A. Al-Awadhi [1], Fahad A. Alammar [1,*], Bader N. Alkhalaf [2] and Hazzaa M. Al-Hazzaa [3]

1 Department of Food and Nutrition Science, College of Health Sciences, The Public Authority for Applied Education & Training, Shuwaikh 70654, Kuwait; ny.bumaryoum@paaet.edu.kw (N.Y.B.); b.alawadhi@paaet.edu.kw (B.A.A.-A.)
2 Department of Environmental Health, College of Health Sciences, The Public Authority for Applied Education & Training, Shuwaikh 70654, Kuwait; bn.alkhalaf@paaet.edu.kw
3 Lifestyle and Health Research Center, Health Sciences Research Center, Princess Nourah Bint Abdulrahman University, Riyadh 11546, Saudi Arabia; halhazzaa@hotmail.com
* Correspondence: fahadaffzalammar@gmail.com

Abstract: The COVID-19 pandemic and the measures implemented to control it have led to widespread lifestyle changes globally. While previous studies have explored these effects across different age groups, this research focuses uniquely on college students in Kuwait. A cross-sectional study (N = 1259) was conducted among college students using a validated online questionnaire covering various aspects, including demographics, academic performance, self-reported body weight and height, sleep duration, dietary habits, and physical activity. There were equal proportions of individuals reporting decreased, increased, or no change in body weight, BMI, and dietary habits due to COVID-19, with no significant gender differences observed. A higher decrease (44.3%) than increase (29.2%) occurred due to COVID-19 in the proportion of college students engaging in physical activity. Significant differences were found in the students' responses to most of the eating behavior questions, concerning the changes attributable to the COVID-19 pandemic restrictions. Multivariable analysis showed significant interaction effects for gender by losing above 9 kg during the last 6 months in age ($p = 0.037$), total physical activity in METs-min/week ($p = 0.048$), and cake/donuts intake ($p = 0.006$). Logistic regression analysis, adjusted for age, revealed that a decrease in BMI was associated with an increase in daily vegetable intake (aOR = 1.531, $p = 0.031$), whereas increases in BMI were associated with an increased intake of sugar-sweetened drinks equal to or above 4 days/week (aOR = 1.551, $p = 0.032$) and increased chocolates/candy intake equal to or above 4 days/week (aOR = 1.792, $p = 0.037$). It was concluded that, in response to a major epidemic, such as COVID-19, college students, as a population, are susceptible to significant changes in lifestyle and eating behaviors that can impact their health and well-being. Future risks for college students' health can be reduced through appropriate lifestyle interventions.

Keywords: COVID-19 pandemic; college students; Kuwait; lifestyle changes; eating habits; physical activity; body weight; sleep patterns; academic achievement; gender differences

1. Introduction

The coronavirus disease 2019 (COVID-19) was declared a pandemic by the World Health Organization in March 2020 [1]. As of 26 August 2023, it had resulted in a staggering global toll of 6.9 million deaths and 770 million reported cases [1]. To mitigate the rapid spread of the virus and its associated morbidity and mortality rates, governments worldwide swiftly implemented a range of measures, including total or partial lockdowns, quarantines, social distancing mandates, and the closure of schools and workplaces [2].

While the impact of such measures was felt across all age groups, children and adolescents emerged as particularly vulnerable to the mental health challenges brought about by the pandemic [3–5]. The disruptions to their usual routines, physical activity, and social interactions, coupled with the stress and uncertainty of the situation, placed students at an increased risk for various psychological issues. It is reasonable to hypothesize that these psychological challenges may have contributed to the changes observed in lifestyle, eating behaviors, and BMI, which are the focus of this research [6]. Numerous studies have reported a surge in negative emotional states among young adults, including heightened levels of anxiety, loneliness, stress, anger, irritability, psychological distress, and depressive symptoms [7,8]. Furthermore, there has been a notable increase in body mass index (BMI) and higher rates of overweight among this demographic [9]. Individuals with pre-existing eating disorders have been especially impacted, with a concerning 41% experiencing a reactivation of their symptoms during lockdowns [8].

Across the globe, the COVID-19 pandemic has influenced many aspects of people's lives such as their health, body weight, education, social life, and lifestyles behaviors [10–12]. Studies conducted in neighboring countries, such as Jordan [9] and Saudi Arabia [13,14], observed similar trends, further emphasizing the regional relevance and importance of this issue. As a consequence of COVID-19, physical activity was impacted with a drastic reduction in exercising, walking, and bicycling for transportation and leisure [15,16]. In addition, as a result of at-home confinement, sedentary behavior frequently replaced time spent in physical activity [16].

College students are generally sociable human beings, and, therefore, the period of social isolation during the COVID-19 lockdowns is considered a major factor that can impact their lifestyle and eating behaviors [12,15,17]. In addition, the closure of universities and the introduction of online teaching definitely disrupted their usual daily routines, added more anxiety and stress to their lives, and may have negatively influenced their mental health and well-being [18]. Hence, it was particularly important to maintain a healthy lifestyle and well-adjusted eating habits during this critical period of the pandemic.

Furthermore, the impacts of COVID-19 on lifestyle behaviors may vary from country to country due to differences in the cultural, social, economic, and types of healthcare systems [11,12,19]. In addition, some recent studies observed gender-related health risk differences as a result of COVID-19 pandemic restrictions [20,21].

In the State of Kuwait, there is limited research examining the impacts of the COVID-19 pandemic on body weight and lifestyle behaviors, including physical activity levels, screen time, sleep, and dietary habits. Therefore, the aim of this study was to evaluate the impact of the COVID-19 pandemic restrictions on self-reported body weight, sleep duration, lifestyle behaviors, eating habits, and academic achievement among college students in Kuwait. Moreover, our study aimed to identify and compare gender differences between male and female students in the measured variables, providing insights into how the pandemic affected each group. Previous research indicated that the pandemic's effects may vary based on gender [20,21].

2. Materials and Methods

2.1. Sample and Design

This cross-sectional study was conducted in Kuwait over a 4-month period from November 2020 to February 2021. Ethical approval was secured from the Institutional Review Board (IRB) at the College of Health Sciences in Kuwait. The sample size was calculated with the assumption that the population proportion would yield the maximum possible sample size required (proportion = 0.50), with a confidence level of 95% and a margin of error of 4%. An additional 20% of participants were added to account for non-responders or missing data. The total sample size for each gender was calculated to be 480, with 960 students in total.

An online questionnaire was answered by college students attending various colleges across the university (under the Public Authority of Applied Education and Training

(PAAET) in Kuwait). The questionnaire was distributed to students electronically, given the restrictions imposed by the pandemic. Participants were provided with clear instructions on how to complete the questionnaire online. The recruitment process involved contacting the colleges' administrations to post the recruitment letter along with the online questionnaire on the colleges' internet sites; teachers who were teaching online were asked by the colleges to advise interested students to participate in the survey. Before completing the questionnaire, students were informed of the study objectives and procedures and consented to participate in the study. The inclusion criteria were college students (males and females), above 17 years old, without having any disease that prevents him/her from freely moving, and that the participant was free of any nutritional disorders.

2.2. Questionnaire and Data Collection

The questionnaire was composed of demographic variables (gender and age,) self-reported body weight (kg) and height (cm), marital status, governorate name, the student's college, the college grade point average (GPA), the high school grade, as well as the lifestyle behaviors items. The lifestyle behaviors questionnaire included questions on physical activity, screen time, sleep duration, and dietary habits. The lifestyle questionnaire was previously validated for adolescents and young people and was previously published [22–25]. The physical activity questionnaire collected extensive data about the time and frequency of engaging in a variety of activities related to light-, moderate-, and vigorous-intensity physical activities in different domains (leisure-type, sports, transportation like walking, and household activities). Then, the activity was assigned a metabolic equivalent of task (MET) based on the Compendium of Physical Activities. Physical activity then was reported either in minutes per week or as activity energy expenditure in METs-min/week.

Body mass index (BMI) was calculated from weight and height data. Students were also asked about their weight and height both before and during the COVID-19 pandemic. This enabled us to assess changes in BMI and body weight during the pandemic period. The current data used in the present study were part of a prior research project [26].

2.3. Statistical Analysis

Data were entered into an SPSS data file, checked, cleaned, and analyzed using the SPSS program, version 22 (IBM, Chicago, IL, USA). Descriptive statistics were calculated and presented as means and standard deviations (SD) or proportions. Cross tabulation with chi-square tests was used to calculate the proportion (%) of changes due to COVID-19 (before versus during the pandemic) in anthropometric measurements, lifestyle factors, and eating behaviors among the college students relative to gender. We also used multivariable analysis to test the differences in the characteristics of those college students who lost 9 kg or more during the last 6 months of COVID-19 relative to gender (the eating-related items contained a question asking about this issue). Bonferroni tests were used for testing between-subjects' effects. In addition, logistic regression analysis was used to test the associations of selected demographic and lifestyle behaviors relative to changes in BMI status due to COVID-19 among the college students, while adjusted for age. An alpha level of ≤ 0.05 was used as the level of significance.

3. Results

The anthropometric characteristics of the participating college students, categorized by gender, are summarized in Table 1. The study included a total of 1259 participants, comprising 54.1% females, with a mean age (SD) of 21.5 (3.1) years. No significant age differences relative to gender were observed ($p = 0.133$). However, significant gender differences were noted in body weight, height, and body mass index (BMI), with males exhibiting higher values ($p < 0.001$). Additionally, more than 10% of the participants were married. The prevalence of overweight or obesity stood at 42%, with males having a significantly higher rate than females ($p < 0.001$).

Table 1. Descriptive characteristics of the participating college students.

Variable	All N = 1259	Male N = 577	Female N = 682	*p*-Value *
Age (year)	21.5 ± 3.1	21.7 ± 3.1	21.4 ± 3.1	0.133
Body weight (kg)	69.5 ± 19.7	78.7 ± 21.2	61.6 ± 14.3	<0.001
Body height (cm)	165.9 ± 9.8	174.1 ± 6.7	158.9 ± 6.0	<0.001
Body mass index (kg/m^2)	25.1 ± 6.3	25.9 ± 7.0	24.3 ± 5.5	<0.001

Data are means ± standard deviations; * *t*-test for independent samples for the differences between males and females.

Table 2 presents the proportions (%) of changes in anthropometric and lifestyle factors among Kuwaiti college students due to COVID-19, categorized by gender. Notably, there were relatively equal proportions of individuals reporting decreased, increased, or no change in body weight, BMI, and dietary habits due to COVID-19, with no significant gender-based differences observed in the proportions of changes (decrease, increase, or no change) in body weight ($p = 0.376$), BMI ($p = 0.417$), or dietary habits ($p = 0.133$). However, more decrease (44.3%) than increase (29.2%) occurred due to COVID-19 in the proportion of Kuwaiti college students engaging in physical activity, with no gender difference ($p = 0.767$). Table 2 provides a detailed breakdown of these changes.

Table 2. Changes (%) in anthropometric and lifestyle factors due to COVID-19 among Kuwaiti college students.

Variable	Change (%) *	All N = 1259	Male N = 577	Female N = 682	*p*-Value **
Body weight	Decrease	34.3	34.7	34.0	0.376
	Increase	27.3	26.2	28.3	
	No change	38.4	39.1	37.7	
BMI	Decrease	34.3	34.7	34.1	0.417
	Increase	27.2	26.1	28.2	
	No change	38.5	39.2	37.7	
Physical activity levels	Decrease	44.3	43.2	45.2	0.767
	Increase	29.2	29.6	28.8	
	No change	26.5	27.2	26.0	
Dietary habits	Deteriorated	27.9	25.8	29.6	0.133
	Improved	38.9	41.8	36.5	
	No change	33.2	32.4	33.9	

* The changes indicate from before to during COVID-19. ** Chi-square tests for the differences in proportions between males and females. Explanation: "Deteriorated" indicates that students reported worsening of dietary habits. "Improved" means students reported better dietary habits. "No Change" implies no significant change in dietary habits.

The proportional changes (%) in eating behaviors among college students resulting from the COVID-19 pandemic and associated restrictions are outlined in Appendix A. As indicated in the table, significant differences were found in the students' responses to most of the eating behavior questions (26 items) concerning the changes attributable to the COVID-19 pandemic restrictions. However, there were no significant differences in the following questions relative to students' responses (not included in Table 3): I feel that those around me would like me to eat more ($p = 0.224$), I throw up after eating foods ($p = 0.574$), I take more time than others when eating my meals ($p = 0.478$), I feel that those around me push me to eat ($p = 0.582$), Foods occupy my time and thinking ($p = 0.358$), I like to have empty stomach ($p = 0.060$), and I feel the urge to throw up after eating meals ($p = 0.962$).

Table 3. Multivariable analysis of the characteristics of those Kuwaiti college students who lost 9 kg or more during the last 6 months of COVID-19 relative to gender (n = 1188).

Variable	Gender	Losing ≥ 9 kg during the Last 6 Months		*p*-Value *
		Never	Yes	
Age (years)	Male Female All	21.4 ± 2.9 21.4 ± 3.2 21.4 ± 3.1	22.1 ± 3.3 21.4 ± 3.0 21.7 ± 3.1	Losing 9 kg: 0.086 Gender: 0.051 Losing 9 kg by gender interaction: 0.037
Body weight (kg)	Male Female All	75.5 ± 20.9 59.9 ± 13.8 66.7 ± 18.9	84.7 ± 20.5 64.5 ± 14.5 74.2 ± 20.3	Losing 9 kg: <0.001 Gender: <0.001 Losing 9 kg by gender interaction: 0.052
BMI (kg/m^2)	Male Female All	25.1 ± 6.5 23.6 ± 5.2 24.3 ± 5.8	27.6 ± 6.4 25.7 ± 5.5 26.6 ± 6.0	Losing 9 kg: <0.001 Gender: <0.001 Losing 9 kg by gender interaction: 0.643
Screen time (hours/day)	Male Female All	4.6 ± 3.1 5.8 ± 3.8 5.3 ± 3.6	4.9 ± 3.7 5.5 ± 4.2 5.2 ± 4.0	Losing 9 kg: 0.988 Gender: <0.001 Losing 9 kg by gender interaction: 0.152
Sleep duration (hours/night)	Male Female All	7.4 ± 1.6 7.4 ± 1.6 7.4 ± 1.6	7.2 ± 1.6 7.5 ± 1.7 7.4 ± 1.6	Losing 9 kg: 0.841 Gender: 0.051 Losing 9 kg by gender interaction: 0.228
Total physical activity (METs-min/week)	Male Female All	2376.7 ± 2935.7 1424.7 ± 1876.2 1840.6 ± 2441.8	2700.9 ± 2832.1 2350.6 ± 2416.4 2519.2 ± 2627.4	Losing 9 kg: <0.001 Gender: <0.001 Losing 9 kg by gender interaction: 0.048
Moderate-intensity physical activity (METs-min/week)	Male Female All	579.4 ± 892.5 782.6 ± 1001.4 693.8 ± 960.1	720.5 ± 1014.6 1118.8 ± 1442.3 927.1 ± 1154.1	Losing 9 kg: <0.001 Gender: <0.001 Losing 9 kg by gender interaction: 0.116
Vigorous-intensity physical activity (METs-min/week)	Male Female All	1797.4 ± 2405.8 642.1 ± 1233.9 1146.8 ± 1926.0	1980.5 ± 2195.4 1231.7 ± 1611.9 1592.1 ± 1949.1	Losing 9 kg: <0.001 Gender: <0.001 Losing 9 kg by gender interaction: 0.078
Breakfast intake (day/week)	Male Female All	4.0 ± 2.7 3.8 ± 2.8 3.9 ± 2.7	3.9 ± 2.7 4.2 ± 2.9 4.0 ± 2.8	Losing 9 kg: 0.671 Gender: 0.799 Losing 9 kg by gender interaction: 0.090
Vegetables intake (day/week)	Male Female All	4.0 ± 2.5 4.0 ± 2.5 4.0 ± 2.5	4.5 ± 2.4 4.4 ± 2.6 4.4 ± 2.5	Losing 9 kg: 0.004 Gender: 0.992 Losing 9 kg by gender interaction: 0.578
Fruit intake (day/week)	Male Female All	3.0 ± 2.3 3.0 ± 2.4 3.0 ± 2.4	3.5 ± 2.5 3.5 ± 2.4 3.5 ± 2.4	Losing 9 kg: <0.001 Gender: 0.913 Losing 9 kg by gender interaction: 0.927
Milk/dairy products intake (day/week)	Male Female All	4.3 ± 2.5 4.2 ± 2.6 4.3 ± 2.5	4.5 ± 2.5 4.3 ± 2.5 4.4 ± 2.5	Losing 9 kg: 0.514 Gender: 0.225 Losing 9 kg by gender interaction: 0.833
Sugar-sweetened drink intake (day/week)	Male Female All	4.0 ± 2.6 3.5 ± 2.5 3.7 ± 2.6	3.7 ± 2.7 3.1 ± 26 3.4 ± 2.7	Losing 9 kg: 0.023 Gender: <0.001 Losing 9 kg by gender interaction: 0.847
Fast food intake (day/week)	Male Female All	3.4 ± 2.1 3.1 ± 2.1 3.2 ± 2.1	3.1 ± 2.1 2.6 ± 1.9 2.8 ± 2.0	Losing 9 kg: 0.001 Gender: <0.001 Losing 9 kg by gender interaction: 0.381
French fries/potato chips intake (day/week)	Male Female All	2.9 ± 2.1 3.1 ± 2.2 3.0 ± 2.1	2.8 ± 2.1 2.5 ± 2.0 2.6 ± 2.1	Losing 9 kg: 0.006 Gender: 0.434 Losing 9 kg by gender interaction: 0.071
Cake/donuts intake (day/week)	Male Female All	2.3 ± 2.1 3.1 ± 2.4 2.7 ± 2.3	2.4 ± 2.1 2.5 ± 2.0 2.5 ± 2.1	Losing 9 kg: 0.101 Gender: 0.001 Losing 9 kg by gender interaction: 0.006

Table 3. *Cont.*

Variable	Gender	Losing ≥ 9 kg during the Last 6 Months		*p*-Value *
		Never	Yes	
Chocolates/candy intake (day/week)	Male Female All	3.2 ± 2.4 3.9 ± 2.4 3.6 ± 2.4	3.0 ± 2.4 3.2 ± 2.3 3.1 ± 2.3	Losing 9 kg: 0.003 Gender: 0.002 Losing 9 kg by gender interaction: 0.074
Energy drinks intake (day/week)	Male Female All	1.5 ± 2.2 1.2 ± 2.1 1.4 ± 2.1	2.0 ± 2.4 1.5 ± 2.3 1.7 ± 2.3	Losing 9 kg: 0.012 Gender: 0.002 Losing 9 kg by gender interaction: 0.364
High school grade (%)	Male Female All	75.1 ± 8.4 75.2 ± 8.9 75.1 ± 8.7	75.4 ± 9.2 73.9 ± 9.1 74.6 ± 9.2	Losing 9 kg: 0.369 Gender: 0.171 Losing 9 kg by gender interaction: 0.146
College GPA	Male Female All	3.00 ± 0.56 3.12 ± 0.61 3.07 ± 0.60	2.95 ± 0.60 2.98 ± 0.61 2.97 ± 0.61	Losing 9 kg: 0.010 Gender: 0.054 Losing 9 kg by gender interaction: 0.279

Data are means and standard deviations. * Wilks' lambda *p* values for multivariable tests: age < 0.001; gender < 0.001; losing 9 kg < 0.001; and gender by losing 9 kg interactions <0.001.

The results of the multivariable analysis, displayed in Table 3, focus on the characteristics of college students who reported losing 9 kg or more during the last 6 months of the COVID-19 pandemic. Significant interaction effects were observed for gender by losing above 9 kg during the last 6 months in age ($p = 0.037$), total physical activity in METs-min/week ($p = 0.048$), and cake/donuts intake ($p = 0.006$). Additionally, the significant, main effects of losing above 9 kg during the last 6 months were noted in several variables including body weight ($p < 0.001$), BMI ($p < 0.001$), total physical activity ($p < 0.001$), moderate-intensity physical activity ($p < 0.001$), vigorous-intensity physical activity ($p < 0.001$), as well as intakes of vegetables ($p = 0.004$), fruits ($p < 0.001$), sugar-sweetened drinks ($p = 0.023$), fast foods ($p = 0.001$), French fries/potato chips ($p = 0.006$), chocolates/candy ($p = 0.003$), and energy drinks ($p = 0.012$). Furthermore, college GPA was also significantly associated with this category ($p = 0.010$).

Table 4 shows the results of logistic regression analysis, adjusted for age, which explores the relationship between selected demographic and lifestyle behaviors and changes in BMI status due to the COVID-19 pandemic among college students. A decrease in BMI of 53% was associated with an increase in daily vegetable intake (aOR = 1.531, $p = 0.031$), whereas increases in BMI of 55% and 79% were associated, respectively, with an increased intake of sugar-sweetened drinks equal to or above 4 days/week (aOR = 1.551, $p = 0.032$) and increased chocolates/candy intake of equal to or above 4 days/week (aOR = 1.792, $p = 0.037$).

Table 4. Logistic regression analysis of lifestyle behaviors relative to changes in BMI status due to COVID-19.

Variable	Changes in BMI Status *			
	aOR	(95% CI)	SEE	*p*-Value
	Decrease in BMI			
Age (younger age = ref)	1.024	0.967–1.084	0.029	0.424
Gender (female = ref)	0.891	0.633–1.253	0.1174	0.506
Marital status (married = ref)	1.257	0.704–2.244	0.296	0.439
Screen time (high screen time = ref)	1.00			
Low screen time	0.990	0.696–1.410	0.180	0.957
Sleep duration (sufficient sleep = ref)	1.00			
Insufficient sleep	0.995	0.702–1.412	0.178	0.979
Physical activity (low active = ref)	1.00			
High active	1.008	0.721–1.409	0.171	0.962
Breakfast intake (non-daily intake = ref)	1.00			

Table 4. *Cont.*

Variable	Changes in BMI Status *			
	aOR	**(95% CI)**	**SEE**	***p*-Value**
Daily intake	1.038	0.723–1.491	0.185	0.838
Vegetables intake (non-daily intake = ref)	1.00			
Daily intake	1.531	1.039–2.258	0.198	0.031
Fruit intake (non-daily intake = ref)	1.00			
Daily intake	1.332	0.832–2.135	0.240	0.233
Milk/dairy products intake (non-daily intake = ref)	1.00			
Daily intake	1.009	0.693–1.467	0.191	0.965
Sugar-sweetened drink intake (<4 days/week = ref)	1.00			
≤ 4 days/week	0.798	0.553–1.150	0.187	0.226
Fast food intake (<4 days/week = ref)	1.00			
≥4 days/week	1.092	0.706–1.689	0.223	0.693
French fries/potato chips intake (<4 days/week = ref)	1.00			
≥4 days/week	0.802	0.513–1.252	0.227	0.331
Cake/donuts intake (<4 days/week = ref)	1.00			
≥4 days/week	0.874	0.515–1.484	0.270	0.618
Chocolates/candy intake (<4 days/week = ref)	1.00			
≥4 days/week	1.247	0.777–2.001	0.241	0.360
Energy drinks intake (<4 days/week = ref)	1.00			
≥4 days/week	0.738	0.453–1.202	0.249	0.222
Increase in BMI				
Age (younger age = ref)	1.011	0.952–1.074	0.031	0.717
Gender (female = ref)	0.855	0.600–1.218	0.181	0.385
Marital status (married = ref)	1.639	0.867–3.097	0.325	0.128
Screen time (high screen time = ref)	1.00			
Low screen time	0.855	0.593–1.233	0.187	0.401
Sleep duration (sufficient sleep = ref)	1.00			
Insufficient sleep	1.184	0.826–1.597	0.184	0.357
Physical activity (low active = ref)	1.00			
High active	0.745	0.524–1.058	0.179	0.100
Breakfast intake (non-daily intake = ref)	1.00			
Daily intake	0.827	0.572–1.196	0.188	0.314
Vegetables intake (non-daily intake = ref)	1.00			
Daily intake	1.124	0.757–1.669	0.202	0.561
Fruit intake (non-daily intake = ref)	1.00			
Daily intake	0.851	0.541–1.339	0.231	0.486
Milk/dairy products intake (non-daily intake = ref)	1.00			
Daily intake	1.414	0.958–2.088	0.199	0.081
Sugar-sweetened drink intake (<4 days/week = ref)	1.00			
≥4 days/week	1.551	1.039–2.314	0.204	0.032
Fast food intake (<4 days/week = ref)	1.00			
≥4 days/week	1.368	0.845–2.216	0.246	0.203
French fries/potato chips intake (<4 days/week = ref)	1.00			
≥4 days/week	0.736	0.455–1.191	0.246	0.212
Cake/donuts intake (<4 days/week = ref)	1.00			
≥4 days/week	0.896	0.501–1.599	0.295	0.710
Chocolates/candy intake (<4 days/week = ref)	1.00			
≥4 days/week	1.792	1.036–3.103	0.280	0.037
Energy drinks intake (<4 days/week = ref)	1.00			
≥4 days/week	0.604	0.365–1.002	0.258	0.051

* No change was used as a reference category; aOR = adjusted odds ratio; CI = confidence interval; ref = reference category; SEE = standard error.

4. Discussion

The COVID-19 pandemic and the subsequent measures, including lockdowns and social distancing, had profound effects on individuals' lifestyles and behaviors across the globe. While previous research predominantly focused on specific age groups such as children, adolescents, and adults, our study uniquely examined the impact of the COVID-19 pandemic on college students' body weight, lifestyle behaviors, and eating habits.

4.1. Impact on Body Weight and BMI

Although nearly one-third of the students in our study experienced an increase in body weight and BMI due to the lockdowns and quarantine measures, the overall findings did not show significant differences among those who increased, decreased, or maintained their body weight. However, the COVID-19 lockdown resulted in weight gain among Jordanian children and adolescents [9,27]. Additionally, among Malaysian university students, there was a weight gain of about 3.4 kg after the first month of the COVID-19 lockdown [17]. Similarly, two systematic reviews and meta-analyses examining nearly 100 studies indicated that during the COVID-19 pandemic lockdowns there were increases amounting to 11.1–72.4% in body weight among young adults in these studies [28,29]. It was believed that the main cause of increased body weight among individuals during COVID-19 lockdowns was the move to online education, which involved higher screen time and long sitting hours along with increased food consumption and snacking [17,28–30]. Also, with lower physical activity levels, lower diet quality, and more frequent overeating, there was a higher chance of increased BMI [31].

4.2. Changes in Lifestyle Behaviors and Eating Habits

The present findings regarding changes in lifestyle behaviors and eating habits align with those observed in various studies conducted during the pandemic [27,32–34]. Furthermore, our research highlights significant changes in eating habits. Increased consumption of unhealthy foods and snacks was prevalent among our participants, which is in line with findings from various international and regional studies [32–34]. Moreover, the findings of the current study appeared consistent with similar findings worldwide indicating that the COVID-19 pandemic had profound effects on lifestyle and eating habits. A study carried out on Portuguese university students showed that healthier eating habits were adopted by the students during the lockdown period compared to a typical semester period. Such healthier behaviors included reduced meal delivery, a lower intake of sugar, salt, and fast food, as well as higher engagement in physical activities and more consumption of fruits, vegetables, and legumes [35]. The decreased physical activity levels observed in our study, especially among females, are consistent with global trends during the pandemic [27,32–34]. Nevertheless, contrary to many other studies' findings, studies from Italy [36] and Portugal [10] found a greater adherence to engaging in physical activity during the lockdown period.

The above changes, due to the COVID-19 lockdown, can be attributed to various factors such as increased stress, boredom, and greater access to unhealthy foods at home [37]. The absence of social interaction, especially among young college students, could lead to increased stress, anxiety, and even depression [18]; such feelings of isolation might elicit emotional eating in order to relieve adverse feelings [38].

Notably, our study indicates that most participants reported changes in eating habits, underscoring the widespread impact of the pandemic on dietary behaviors. Lockdown can bring about some healthy eating behaviors. Eating habits of college students from Tabriz (Iran) were characterized by an increase in a regular meal pattern and balanced diet, milk or its products, and egg or meat in the diet, as well as by a reduction in fast food intake, fried food, potato chips, and sugar-sweetened drinks. However, the students reported increases in weight gain, screen and sitting time, and sleep duration, as well as declined physical activity and higher stress or anxiety levels during the COVID-19 pandemic [37]. Similar studies carried out in Italy, Spain, Romania, and Latin America showed some improvement in dietary habits including an increased intake of fruits, vegetables, yogurt, legumes, and eggs during the lockdown period [36,39–41].

The findings of the present study do not agree with results found in a study conducted on adults from the UK, which showed a higher number of participants reporting more negative changes in their eating habits, including more frequent snacking, during the COVID-19 confinement compared to the pre-lockdown period [30]. In addition, a Portuguese study, conducted on adults, observed that the majority of the participants consumed more foods rich in fat and salt and sugar-sweetened beverages during the restriction period compared to pre-pandemic [10].

4.3. Sleep Patterns and Academic Achievement

Sleep duration, an essential component of overall health, also experienced alterations during the pandemic. We observed that a substantial number of students reported shifts in their sleep patterns, with many experiencing later bedtimes and wake times. This pattern mirrors the changes in daily routines brought about by remote learning and reduced social commitments, as seen in other studies [32,33]. Despite that, our findings did not show any significant changes in sleep duration relative to a decrease or increase in weight or BMI. However, the interaction of sleep duration, BMI, and lifestyle factors is quite complex, as exhibited in our current study.

Our findings on academic achievement reveal a mixed picture. While some students reported improvements in academic performance due to the flexibility of online learning, others experienced challenges related to motivation and concentration. These variations may be attributed to individual differences in adapting to remote learning environments as well as the overall impact of the pandemic on mental health and stress levels [7,8,34].

4.4. Gender Differences

Aside from the pandemic, it is well recognized that women on average report more physical and mental unhealthy days per year than men [42]. However, reports indicate that women are at a reduced risk of severe disease and death with SARS-CoV-2 infection than men [43,44]. Gender differences in our study show that there were greater changes among females in their body weight, physical activity, and sleep patterns. These gender differences align with the observations made by Al Hourani et al., who noted significant increases in body weight among females [27]. In addition, recent studies observed some differences in gender-related health risks resulting from COVID-19 pandemic restrictions [20,21]. Further, the psychological abuse of women as a type of violence reportedly increased significantly during the lockdown [45].

4.5. Strengths, Limitations, and Future Research

The current research strengths included a large sample of male and female college students from a variety of colleges and the use of a validated lifestyle questionnaire. However, our study is not without limitations. One notable limitation is the use of self-reported data for variables such as body weight and height, which may introduce reporting bias due to a social desirability effect. Additionally, there is a small chance that the sample may not be fully representative of all Kuwaiti college students, although we targeted all college students when recruiting participants.

Future research could involve more objective measures, such as clinical assessments of body composition and the use of accelerometers for the assessment of physical activity, sedentary behaviors, and sleep, to mitigate the limitations associated with self-reported data. Longitudinal studies are also warranted to assess the long-term consequences of the observed lifestyle changes and their impact on the health and well-being of college students.

5. Conclusions

As the result of a major epidemic, such as COVID-19 restrictions, college students appear to be susceptible to significant changes in lifestyle and eating behaviors that can impact their health and well-being. To the best of our knowledge, the current research is the first study that reported about the effects of the COVID-19 pandemic on body weight, lifestyle behaviors, and eating habits among college students in the State of Kuwait using a fairly large and representative sample. The COVID-19 pandemic has profoundly impacted the lives of college students in Kuwait, manifesting in significant changes in anthropometry, lifestyle behaviors, and dietary habits. This study, despite its local focus, contributes to the growing body of literature on the pandemic's effects, emphasizing the need for tailored interventions and ongoing research. The pandemic's influence on body weight and BMI is evident, with nearly 30% of college students showing an increase in body weight and BMI. Reduced physical activity levels, dietary changes, disrupted sleep patterns, and variations in academic performance further underscore the comprehensive impact.

Future risks for college students' health can be reduced through appropriate lifestyle interventions, hence, promoting healthier lifestyle behaviors and dietary habits among college students during crises. These interventions should be multidimensional, addressing not only the physical health but also the psychosocial well-being of students. While our research provides valuable insights into the specific context of Kuwaiti college students, more extensive investigations are warranted to explore the underlying mechanisms and long-term implications of these changes. Collaboration among policymakers, educators, healthcare professionals, and parents is essential to foster a healthier environment for college students amid and beyond the pandemic.

Author Contributions: Conceptualization, A.R.A.-H., H.M.A.-H. and N.Y.B.; methodology, A.R.A.-H., H.M.A.-H. and B.A.A.-A.; software, H.M.A.-H. and B.N.A.; validation, A.R.A.-H., N.Y.B. and B.N.A.; formal analysis, B.A.A.-A. and F.A.A.; investigation, A.R.A.-H., N.Y.B., F.A.A. and B.N.A.; resources, N.Y.B. and B.N.A.; data curation, H.M.A.-H. and B.N.A.; writing—original draft preparation, H.M.A.-H. and F.A.A.; writing—review and editing, A.R.A.-H. and F.A.A.; visualization, A.R.A.-H., H.M.A.-H., N.Y.B. and B.N.A.; supervision, A.R.A.-H. and N.Y.B.; project administration, A.R.A.-H., B.N.A. and N.Y.B. All authors have read and agreed to the published version of the manuscript.

Funding: This research received no external funding.

Institutional Review Board Statement: The Ministry of Education and the Institutional Review Board (IRB) at the College of Health Sciences in Kuwait approved the study protocol (Code: 1-2/2020; 10 September 2020) before starting data collection.

Informed Consent Statement: The Ministry of Education and the Institutional Review Board (IRB) at the College of Health Sciences in Kuwait approved the study protocol before starting data collection. Dr. Ahmad and contributing authors give consent for the above manuscript to be published in Nutrients Special Issue "Body Weight and Food/Eating-Related Behaviors during COVID-19 Pandemic, Traumatic or Stressful Life Events".

Data Availability Statement: Data are contained within the article.

Acknowledgments: We would like to thank all the participants for taking part in this study.

Conflicts of Interest: The authors declare no conflict of interest. The funders had no role in the design of the study; in the collection, analyses, or interpretation of data; in the writing of the manuscript; or in the decision to publish the results.

Appendix A

Table A1. Responses to questions related to eating behaviors as a result of changes in body weight due to COVID-19 among Kuwaiti college students.

Variable	Response Choice	Change in Weight Due to COVID-19 (%)			p-Value
		Decreased N = 432	Increased N = 344	No Change N = 483	
I am scared of being overweight	Never	12.5	16.3	29.4	<0.001
	Rarely	6.5	8.4	14.7	
	Occasionally	19.2	16.0	14.3	
	Mostly	22.0	24.7	17.6	
	Always	39.8	34.6	24.0	

Table A1. *Cont.*

Variable	Response Choice	Change in Weight Due to COVID-19 (%)			p-Value
		Decreased N = 432	Increased N = 344	No Change N = 483	
I avoid eating if I am hungry	Never	27.8	18.6	31.1	0.001
	Rarely	20.8	17.2	20.1	
	Occasionally	26.9	34.6	22.8	
	Mostly	19.7	23.3	21.7	
	Always	4.9	6.4	4.3	
I have difficulty stopping eating	Never	26.6	35.8	49.1	<0.001
	Rarely	17.8	25.9	17.4	
	Occasionally	23.8	17.7	15.1	
	Mostly	15.8	13.1	13.0	
	Always	13.2	7.6	5.4	
I eat my meals as several small snacks	Never	39.8	28.5	31.7	0.005
	Rarely	18.3	18.9	17.8	
	Occasionally	21.1	20.1	23.4	
	Mostly	12.3	19.5	18.4	
	Always	8.6	13.1	8.7	
I know the calories of my eaten foods	Never	51.6	31.1	49.3	<0.001
	Rarely	14.4	14.0	12.6	
	Occasionally	13.9	18.9	12.8	
	Mostly	13.0	19.2	15.9	
	Always	7.2	16.9	9.3	
I avoid carbohydrate-rich foods (like bread, potato, and rice)	Never	42.4	33.4	45.3	0.003
	Rarely	16.9	16.0	14.7	
	Occasionally	20.6	20.3	15.9	
	Mostly	14.6	19.8	18.4	
	Always	5.6	10.5	5.6	
I feel guilty after eating	Never	32.2	32.6	55.1	<0.001
	Rarely	13.4	13.4	12.0	
	Occasionally	18.8	22.7	11.0	
	Mostly	16.0	17.2	12.2	
	Always	19.7	14.2	9.7	
I am obsessed with the desire to be thin	Never	16.4	20.3	44.3	<0.010
	Rarely	6.5	7.6	8.3	
	Occasionally	13.7	13.7	11.6	
	Mostly	15.0	17.7	13.9	
	Always	48.4	40.7	21.9	
I think about burning calories when doing exercise	Never	14.6	16.9	32.5	<0.001
	Rarely	7.9	4.4	6.6	
	Occasionally	12.0	11.0	13.5	
	Mostly	24.5	21.2	20.3	
	Always	41.0	46.5	27.1	
People think I am very skinny	Never	35.9	31.1	26.3	0.035
	Rarely	10.6	12.8	9.9	
	Occasionally	16.9	16.0	15.9	
	Mostly	19.0	19.8	22.6	
	Always	17.6	20.3	25.3	
I am obsessed with the idea that I have fats in my body	Never	14.1	15.7	30.6	<0.001
	Rarely	9.0	6.1	10.8	
	Occasionally	15.5	18.3	16.6	
	Mostly	22.2	23.3	18.2	
	Always	39.1	36.6	23.8	

Table A1. *Cont.*

Variable	Response Choice	Change in Weight Due to COVID-19 (%)			p-Value
		Decreased N = 432	Increased N = 344	No Change N = 483	
I avoid foods with sugar	Never	29.2	18.0	33.5	<0.001
	Rarely	18.8	13.1	17.2	
	Occasionally	27.8	27.6	23.0	
	Mostly	17.6	25.6	19.5	
	Always	6.7	15.7	6.8	
I eat diet foods	Never	39.1	28.2	41.4	<0.001
	Rarely	16.7	14.5	16.8	
	Occasionally	24.1	21.2	19.0	
	Mostly	16.2	23.3	16.8	
	Always	3.9	12.8	6.0	
I feel that diet is controlling my life	Never	26.4	26.5	37.2	0.001
	Rarely	10.9	11.6	12.2	
	Occasionally	17.4	20.1	16.8	
	Mostly	25.2	19.5	19.9	
	Always	20.1	22.4	14.1	
I can control myself in front of foods	Never	20.8	11.9	24.2	0.001
	Rarely	11.1	11.9	10.8	
	Occasionally	25.0	21.5	19.7	
	Mostly	26.9	31.7	24.6	
	Always	16.2	23.0	20.7	
I do not feel comfortable after eating sweets	Never	30.3	27.0	34.2	0.007
	Rarely	14.6	13.4	15.3	
	Occasionally	22.0	20.3	18.2	
	Mostly	13.7	20.6	20.1	
	Always	19.4	18.6	12.2	
I practice certain dietary behaviors	Never	43.1	30.8	51.1	<0.001
	Rarely	14.6	13.1	12.2	
	Occasionally	19.2	15.7	14.1	
	Mostly	16.0	24.1	14.1	
	Always	7.2	16.3	8.5	
I enjoy eating new fatty foods	Never	5.2	5.6	6.8	0.020
	Rarely	16.0	21.8	14.7	
	Occasionally	21.5	21.8	25.5	
	Mostly	25.7	21.2	22.2	
	Always	21.8	14.8	20.5	

References

1. World Health Organization. WHO COVID-19 Dashboard. [Online] World Health Organization. 2023. Available online: https://covid19.who.int (accessed on 7 August 2023).
2. Haug, N.; Geyrhofer, L.; Londei, A.; Dervic, E.; Desvars-Larrive, A.; Loreto, V.; Pinior, B.; Thurner, S.; Klimek, P. Ranking the effectiveness of worldwide COVID-19 government interventions. *Nat. Hum. Behav.* **2020**, *4*, 1303–1312. [CrossRef] [PubMed]
3. Guessoum, S.B.; Lachal, J.; Radjack, R.; Carretier, E.; Minassian, S.; Benoit, L.; Moro, M.R. Adolescent Psychiatric Disorders during the COVID-19 Pandemic and Lockdown. *Psychiatry Res.* **2020**, *291*, 113264. [CrossRef] [PubMed]
4. Wang, G.; Zhang, Y.; Zhao, J.; Zhang, J.; Jiang, F. Mitigate the Effects of Home Confinement on Children during the COVID-19 Outbreak. *Lancet* **2020**, *395*, 945–947. [CrossRef] [PubMed]
5. Marciano, L.; Ostroumova, M.; Schulz Peter, J.; Camerini, A.L. Digital Media Use and Adolescents' Mental Health during the COVID-19 Pandemic: A Systematic Review and Meta-Analysis. *Front. Public Health* **2022**, *9*, 793868. [CrossRef]
6. Chaabane, S.; Doraiswamy, S.; Chaabna, K.; Mamtani, R.; Cheema, S. The Impact of COVID-19 School Closure on Child and Adolescent Health: A Rapid Systematic Review. *Children* **2021**, *8*, 415. [CrossRef]
7. Panchal, U.; de Pablo, G.S.; Franco, M.; Moreno, C.; Parellada, M.; Arango, C.; Fusar-Poli, P. The Impact of COVID-19 Lockdown on Child and Adolescent Mental health: Systematic Review. *Eur. Child. Adolesc. Psychiatry* **2023**, *32*, 1151–1177. [CrossRef]

8. Cielo, F.; Ulberg, R.; Di Giacomo, D. Psychological Impact of the COVID-19 Outbreak on Mental Health Outcomes among Youth: A Rapid Narrative Review. *Int. J. Environ. Res. Public Health* **2021**, *18*, 6067. [CrossRef]
9. Hammouri, H.; Almomani, F.; Abdel Muhsen, R.; Abughazzi, A.; Daghmash, R.; Abudayah, A.; Hasan, I.; Alzein, E. Lifestyle Variations during and after the COVID-19 Pandemic: A Cross-Sectional Study of Diet, Physical Activities, and Weight Gain among the Jordanian Adult Population. *Int. J. Environ. Res. Public Health* **2022**, *19*, 1346. [CrossRef]
10. Antunes, R.; Frontini, R.; Amaro, N.; Salvador, R.; Matos, R.; Morouço, P.; Rebelo-Gonçalves, R. Exploring lifestyle habits, physical activity, anxiety and basic psychological needs in a sample of Portuguese adults during COVID-19. *Int. J. Environ. Res. Public Health* **2020**, *17*, 4360. [CrossRef]
11. Pietrobelli, A.; Pecoraro, L.; Ferruzzi, A.; Heo, M.; Faith, M.; Zoller, T.; Antoniazzi, F.; Piacentini, G.; Fearnbach, S.N.; Heymsfield, S.B. Effects of COVID-19 lockdown on lifestyle behaviors in children with obesity living in Verona, Italy: A longitudinal study. *Obesity* **2020**, *28*, 1382–1385. [CrossRef]
12. Xiang, M.; Zhang, Z.; Kuwahara, K. Impact of COVID-19 pandemic on children and adolescents' lifestyle behavior larger than expected. *Prog. Cardiovasc. Dis.* **2020**, *63*, 531. [CrossRef] [PubMed]
13. Bakhsh, M.A.; Khawandanah, J.; Naaman, R.K.; Alashmali, S. The Impact of COVID-19 Quarantine on Dietary Habits and Physical Activity in Saudi Arabia: A cross-sectional Study. *BMC Public Health* **2021**, *21*, 1487. [CrossRef] [PubMed]
14. Abdulsalam, N.M.; Khateeb, N.A.; Aljerbi, S.S.; Alqumayzi, W.M.; Balubaid, S.S.; Almarghlani, A.A.; Ayad, A.A.; Williams, L.L. Assessment of Dietary Habits and Physical Activity Changes during the Full COVID-19 Curfew Period and Its Effect on Weight among Adults in Jeddah, Saudi Arabia. *Int. J. Environ. Res. Public Health* **2021**, *18*, 8580. [CrossRef] [PubMed]
15. Park, A.H.; Zhong, S.; Yang, H.; Jeong, J.; Lee, C. Impact of COVID-19 on physical activity: A rapid review. *J. Glob. Health.* **2022**, *12*, 05003. [CrossRef]
16. Evenson, K.R.; Alothman, S.A.; Moore, C.C.; Hamza, M.M.; Rakic, S.; Alsukait, R.F.; Herbst, C.H.; Baattaiah, B.A.; AlAhmed, R.; Al-Hazzaa, H.M.; et al. A scoping review on the impact of the COVID-19 pandemic on physical activity and sedentary behavior in Saudi Arabia. *BMC Public Health* **2023**, *23*, 572. [CrossRef]
17. Chen, H.W.J.; Marzo, R.R.; Anton, H.; Abdalqader, M.A.; Rajasekharan, V.; Baobaid, M.F.; Hamzah, H.; Tang, H.C.; Ads, H.O. Dietary habits, shopping behavior and weight gain during COVID-19 pandemic lockdown among students in a private university in Selangor, Malaysia. *J. Public Health Res.* **2021**, *10* (Suppl. S2), jphr.2021.2921. [CrossRef]
18. Verma, K. The mental health impact of the COVID-19 epidemic on college students in India. *Asian J. Psychiatry* **2020**, *53*, 102398. [CrossRef]
19. Chopra, S.; Ranjan, P.; Singh, V.; Kumar, S.; Arora, M.; Hasan, M.S.; Kasiraj, R.; Kaur, D.; Vikram, N.K.; Malhotra, A.; et al. Impact of COVID-19 on lifestyle-related behaviours—A cross-sectional audit of responses from nine hundred and ninety-five participants from India. *Diabetes Metab. Syndr.* **2020**, *14*, 2021–2030. [CrossRef]
20. Schokkenbroek, J.M.; Hardyns, W.; Anrijs, S.; Ponnet, K. Partners in lockdown: Relationship stress in men and women during the COVID-19 pandemic. *Couple Fam. Psychol. Res. Pract.* **2021**, *10*, 149. [CrossRef]
21. Connor, J.; Madhavan, S.; Mokashi, M.; Amanuel, H.; Johnson, N.R.; Pace, L.E.; Bartz, D. Health risks and outcomes that disproportionately affect women during the COVID-19 pandemic: A review. *Soc. Sci. Med.* **2020**, *1*, 113364. [CrossRef]
22. Al-Hazzaa, H.M.; Musaiger, A.O.; ATLS Research Group. Arab Teens Lifestyle Study (ATLS): Objectives, design, methodology and implications. *Diabetes Metab. Syndr. Obes.* **2011**, *4*, 417–426. [CrossRef] [PubMed]
23. Al-Hazzaa, H.M.; Abahussain, N.A.; Al-Sobayel, H.I.; Qahwaji, D.M.; Musaiger, A.O. Physical activity, sedentary behaviors and dietary habits among Saudi adolescents relative to age, gender and region. *Int. J. Behav. Nutr. Phys. Act.* **2011**, *8*, 140. [CrossRef] [PubMed]
24. Al-Hazzaa, H.; Al-Ahmadi, M. A self-reported questionnaire for the assessment of physical activity in youth 15–25 years: Development, reliability and construct validity. *Arab. J. Food Nutr.* **2003**, *4*, 279–291.
25. Al-Hazzaa, H.M.; Al-Sobayel, H.I.; Musaiger, A.O. Convergent validity of the Arab Teens Lifestyle Study (ATLS) physical activity questionnaire. *Int. J. Environ. Res. Public Health* **2011**, *8*, 3810–3820. [CrossRef]
26. Al-Haifi, A.R.; Al-Awadhi, B.A.; Bumaryoum, N.Y.; Alajmi, F.A.; Ashkanani, R.H.; Al-Hazzaa, H.M. The association between academic performance indicators and lifestyle behaviors among Kuwaiti college students. *J. Health Popul. Nutr.* **2023**, *42*, 27. [CrossRef]
27. Al Hourani, H.; Alkhatib, B.; Abdullah, M. Impact of COVID-19 Lockdown on Body weight, Eating Habits and Physical Activity of Jordanian Children and Adolescents. *Disaster Med. Public. Health Prep.* **2022**, *16*, 1855–1863. [CrossRef]
28. Bakaloudi, D.R.; Barazzoni, R.; Bischoff, S.C.; Breda, J.; Wickramasinghe, K.; Chourdakis, M. Impact of the first COVID-19 lockdown on body weight: A combined systematic review and a meta-analysis. *Clin. Nutr.* **2022**, *41*, 3046–3054. [CrossRef]
29. Chang, T.H.; Chen, Y.C.; Chen, W.Y.; Chen, C.Y.; Hsu, W.Y.; Chou, Y.; Chang, Y.H. Weight gain associated with COVID-19 lockdown in children and adolescents: A systematic review and meta-analysis. *Nutrients* **2021**, *13*, 3668. [CrossRef]
30. Rafique, N. Body mass index, eating habits, and various lifestyle changes in young adults during the two years of the coronavirus disease 2019 (COVID-19) pandemic. *J. Fam. Commun. Med.* **2022**, *29*, 189–195.
31. Robinson, E.; Boyland, E.; Chisholm, A.; Harrold, J.; Maloney, N.G.; Marty, L.; Mead, B.R.; Noonan, R.; Hardman, C.A. Obesity, eating behavior and physical activity during COVID-19 lockdown: A study of UK adults. *Appetite* **2021**, *156*, 104853. [CrossRef]
32. Hashem, S.A.; Refay, A.S.E.; Mostafa, H.H.; Kamel, I.H.; Sherif, L.S. Impact of Coronavirus Disease-19 Lockdown on Egyptian Children and Adolescents: Dietary Pattern Changes Health Risk. *Open Access Maced. J. Med. Sci.* **2020**, *8*, 561–569. [CrossRef]

33. Al-Mulla, N.A.; Mahfoud, Z.R. The Impact of the COVID-19 Lockdown 'Home Quarantine' on the Physical Activity and Lifestyle of Children in Qatar. *Front Public Health* **2022**, *10*, 877424. [CrossRef] [PubMed]
34. Alamri, E.S. Effects of COVID-19 Home Confinement on Eating behavior: A Review. *J. Public Health Res.* **2021**, *10*, 2088. [CrossRef] [PubMed]
35. Monteiro, M.; Ferreira-Pêgo, C. University students eating habits: Normal semester vs. lockdown period caused by COVID-19 pandemic. *Int. J. Environ. Res. Public Health* **2022**, *19*, 12750. [CrossRef]
36. Di Renzo, L.; Gualtieri, P.; Pivari, F.; Soldati, L.; Attinà, A.; Cinelli, G.; Leggeri, C.; Caparello, G.; Barrea, L.; Scerbo, F.; et al. Eating habits and lifestyle changes during COVID-19 lockdown: An Italian survey. *J. Transl. Med.* **2020**, *18*, 1–15. [CrossRef]
37. Rafraf, M.; Molani-Gol, R.; Sahebjam, M. Effect of COVID-19 pandemic on eating habits and lifestyle of college students in Tabriz, Iran: A cross-sectional study. *Front. Public Health* **2023**, *11*, 1185681. [CrossRef]
38. Pop, C.; Ciomag, V. Impact of COVID-19 lockdown on body mass index in young adults. *Phys. Educ. Stud.* **2021**, *25*, 98–102. [CrossRef]
39. Celorio-Sardà, R.; Comas-Basté, O.; Latorre-Moratalla, M.L.; Zerón-Rugerio, M.F.; Urpi-Sarda, M.; Illán-Villanueva, M.; Farran-Codina, A.; Izquierdo-Pulido, M.; Vidal-Carou, M.D.C. Effect of COVID-19 lockdown on dietary habits and lifestyle of food science students and professionals from Spain. *Nutrients* **2021**, *13*, 1494. [CrossRef]
40. Enriquez-Martinez, O.G.; Martins, M.C.; Pereira, T.S.; Pacheco, S.O.; Pacheco, F.J.; Lopez, K.V.; Huancahuire-Vega, S.; Silva, D.A.; Mora-Urda, A.I.; Rodriguez-Vásquez, M.; et al. Diet and lifestyle changes during the COVID-19 pandemic in Ibero-American countries: Argentina, Brazil, Mexico, Peru, and Spain. *Front. Nutr.* **2021**, *8*, 671004. [CrossRef]
41. Năstăsescu, V.; Mititelu, M.; Stanciu, T.I.; Drăgănescu, D.; Grigore, N.D.; Udeanu, D.I.; Stanciu, G.; Neacșu, S.M.; Dinu-Pîrvu, C.E.; Oprea, E.; et al. Habits and lifestyle of Romanians in the context of the COVID-19 pandemic. *Nutrients* **2022**, *14*, 504. [CrossRef]
42. Centers for Disease Control and Prevention CDC health disparities and inequalities report—United States, 2013. *Morb. Mortal. Wkly. Rep.* **2013**, *62* (Suppl. S3), 85–154.
43. Purdie, A.; Hawkes, S.; Buse, K.; Onarheim, K.; Aftab, W.; Low, N.; Tanaka, S. Sex, gender and COVID-19: Disaggregated data and health disparities. *BMJ Glob. Health Blog.* **2020**, *39*, 55–73. Available online: https://blogs.bmj.com/bmjgh/2020/03/24/sex-gender-and-covid-19-disaggregated-data-and-health-disparities/ (accessed on 12 August 2023).
44. Richardson, S.; Hirsch, J.S.; Narasimhan, M.; Crawford, J.M.; McGinn, T.; Davidson, K.W.; Barnaby, D.P.; Becker, L.B.; Chelico, J.D.; Cohen, S.L.; et al. Presenting characteristics, comorbidities, and outcomes among 5700 patients hospitalized with COVID-19 in the New York city area. *J. Am. Med. Assoc.* **2020**, *323*, 2052–2059. [CrossRef] [PubMed]
45. Sediri, S.; Zgueb, Y.; Ouanes, S.; Ouali, U.; Bourgou, S.; Jomli, R.; Nacef, F. Women's mental health: Acute impact of COVID-19 pandemic on domestic violence. *Arch. Womens Ment. Health* **2020**, *23*, 749–756. [CrossRef] [PubMed]

nutrients

MDPI

Article

Stress during Home Confinement Is Associated with Eating Misalignment among Adults during COVID-19 Lockdown

Abeer M. Aljaadi *, Rinal J. Bogis, Nouf A. Alruhili, Saja O. Alharbi and Essra A. Noorwali

Clinical Nutrition Department, Faculty of Applied Medical Sciences, Umm Al-Qura University, Makkah 24381, Saudi Arabia; eanoorwali@uqu.edu.sa (E.A.N.)
* Correspondence: amjaadi@uqu.edu.sa

Abstract: Background: The COVID-19 pandemic forced Saudi Arabia to implement several measures including mandatory home confinement, banning entry to many cities, and suspending religious activities. Studies have reported inconsistent findings of the effect of home confinement on lifestyle factors. This study aims to assess the psychological impact of COVID-19 during home confinement and explore its association with dietary habits and weight change. Methods: A cross-sectional study was conducted among Saudi adults using an online survey between May and June 2020. Data on dietary habits, sleep quality, and stress were collected. Results: A total of n = 503 participants responded. Of 254 analyzed, 87% were females, 49% were overweight/obese (body mass index (BMI) $\geq$ 25 kg/m^2), and 79% were under lockdown for >40 days. In multiple linear regression, higher stress scores during confinement were associated with higher stress scores before confinement and poorer sleep quality. In multiple logistic regression, those who did not eat at the same time had higher stress scores compared to those who always ate at the same time, whereas consuming $\geq$three meals was associated with lower stress scores than consuming one–two meals. The odds of gaining weight during confinement were higher among married adults, those with lower sleep quality, and consuming $\geq$three meals. Conclusions: Stress during home confinement was associated with eating misalignment and the number of meals consumed. Although this study was limited by its cross-sectional design and reliance on self-reported data, it provides valuable insights into the dietary habits and weight-gain associated factors that need to be further explored and addressed in any future restrictions for improved well-being.

Keywords: COVID-19; Saudi Arabia; dietary habits; eating misalignment; weight; home confinement; curfew; lockdown; stress; sleep quality

1. Introduction

On 11 March 2020, the World Health Organization designated the coronavirus disease of 2019 (COVID-19) emergency to be a global pandemic [1]. Many governments of countries that were extensively affected implemented policies that restricted social communication to limit COVID-19 spread, including Saudi Arabia [2]. Saudi Arabia implemented a range of measures including a curfew, mandatory home confinement, banning entry to many cities, and suspending religious activities starting 9 March 2020 [2]. Despite the success in controlling COVID-19 spread, the mandated restrictions adversely affected lifestyle behaviors, such as poor sleep quality, poor dietary habits, socioemotional disorders, weight gain, increased screen time, and lack of physical activity in several countries [3–6]. Therefore, studying the impact of home confinement on lifestyle behaviors is essential to prevent adverse changes in future pandemics.

Several studies reported mental health problems during the COVID-19 pandemic [7,8]. The main mental health stressors during the quarantine included the duration of the quarantine, fears of infection, frustration and boredom, inadequate supplies, and inadequate information [9]. Stress can lead to health problems including an increase in blood pressure, blood glucose, and body weight. Studies have shown that persistent stress weakens the immune system via increasing glucocorticoids and pro-inflammatory biomarkers, such as interleukin-6 [10]. Higher cortisol levels are associated with a greater consumption of foods, and thus weight gain [11]. Consequently, assessing the impact of stress on dietary behaviors during home confinement needs to be addressed.

Several studies have demonstrated heterogeneity in eating behaviors and weight change in response to stress; some people eat more when stressed while others eat less. Moreover, different factors during COVID-19 lockdowns could influence the relationship between stress and dietary habits or weight gain, including alterations in physical activity, screen time, sleep quality, and employment status. Higher stress has been generally associated with increased caloric intake, snacking, and consumption of high fat/sugar foods [12]. However, the consumption of homemade meals and breakfast throughout the pandemic increased, and the percentage of those eating mostly fast food and skipping meals decreased [13–17]. Whether dietary changes during the COVID-19 pandemic were stress-related is not clearly understood. Individuals who reported restricted eating, skipping meals, or overeating during the pandemic reported having higher stress levels, after adjusting for sex, age, race, ethnicity, employment, marital status, the region in the US, and BMI [18]; sleep quality during the outbreak and stress level before confinement were not accounted for, which could influence the relationship [16,19].

Increased sitting and screen time, and exposure to stress related to lockdown were correlated with weight increase [17,20], but these studies did not control for potential confounders. Perceived stress scores during the COVID-19 outbreak were not associated with weight increase among women in Riyadh, after adjusting for age, education, sleep quality, physical activity, and baseline stress [16]. A study on U.S. adults reported that weight gain did not correlate with perceived stress scores in bivariate analysis [21]. Some studies have reported better sleep, increased physical activity, and improved diet quality during the pandemic [4,13,14], while others reported worsened lifestyle changes. According to a study conducted on 838 U.S. adults during the last week of April 2020, mean stress scores, as measured by the perceived stress scale, varied by ethnic groups, were higher among people who had part-time jobs, were single, and were younger than 35 years [18]. Inconsistent results between studies could be due to the lack of controlling for such confounding factors, and more research is needed to explore the relationship between stress and changes in dietary habits and weight status. Therefore, the present study aimed to explore the psychological impact of COVID-19 home confinement on dietary habits and weight change taking into consideration changes in sleep quality and physical activity.

2. Materials and Methods

2.1. Study Design

This was a cross-sectional study conducted during the COVID-19 lockdown in Saudi Arabia. Restrictions during COVID-19 were updated almost weekly, with the months of mid-March to May 2020 involving a 24-h curfew and forcing people into home confinement. During home confinement, individuals were not allowed to leave their houses except with an electronic permit [2]. Permit types differed, but all were obtained through a mobile app (Tawakkalna) [22]. Healthcare workers and some employees had exceptions during these lockdowns; however, the majority of the population was under home confinement. Social distancing measures included keeping a distance of at least 1.5 m between people, prohibiting all public gatherings (including religious activities), limiting attendance at private events like weddings and funerals to five people, and forbidding meetings with more than one person who is not a member of one's immediate family. As part of the lockdown procedures, eating establishments, a number of retail stores, and access to outdoor parks were all prohibited.

A 118-item online survey was administered using Google Form® in Arabic and English languages, which included information on sociodemographic, home confinement, health, sleep quality, stress, physical activity, and eating habits. Several questions were presented in a distinguishing formula with questions that focused on change that occurred "during" compared to "before" the confinement period. Responses after June 2020 were excluded, as several restrictions were lifted on 21 June. Only adults (>18 years) who were living in Saudi Arabia at the time of the survey were included. Pregnant and lactating women were excluded because of their altered stress level, sleep quality, and dietary habits [23–25].

2.2. *Dietary Habits and Physical Activity*

A few questions on dietary habits were administered, such as the usual number of main meals consumed, usual number of snacks consumed, frequency of eating at restaurants or take-away or delivered food, frequency of having meals at the same time, and weight change during confinement compared to before confinement. Physical activity was assessed using three questions from the short form of the International Physical Activity Questionnaire [26]. A question on screen time was included by asking about the duration spent in front of screens (phone, computer, TV, etc.) before and during confinement. Body weight and height were self-reported and Body Mass Index (BMI) was calculated as weight (kg)/height $(m)^2$. BMI was classified into 3 categories: underweight (<18.5 kg/m^2), healthy weight (18.5–24.9 kg/m^2), and overweight/obese ($\geq$25 kg/m^2) [27].

2.3. *Sleep Quality and Stress Assessment*

Sleep quality was assessed using the Pittsburgh Sleep Quality Index (PSQI). Each element was given a score between 0 and 3, with a total score ranging from 0 to 21. Higher scores indicate disturbed sleep. Perceived sleep quality was assessed using the Pittsburgh Sleep Quality Index (PSQI). Each element was given a score between 0 and 3, with a total score ranging from 0 to 21. Higher scores indicate disturbed sleep [28]. Questions on stress were adopted from the depression subscale in the Calgary Symptoms of Stress Inventory (C-SOSI). The C-SOSI includes 56-item scale and 8 domains (sub-scales). One of the sub-scales, relating to the variety of emotions that accompany depression, was used as an indicator of affective stress responding to home confinement [29]. Questions on the subscale were formatted to "before" and "during" confinement. Response options were as follows: 0 = never, 1 = almost never, 2 = sometimes, 3 = most of the time, 4 = almost always. The total sum of the subscale score was used with higher scores indicating higher stress levels.

2.4. *Statistical Analyses*

Descriptive statistics were used to define means SD, medians [25th, 75th percentiles], or proportions. To assess differences in continuous variables before and during the confinement period, paired sample *t*-tests or Wilcoxon test were used as appropriate. McNemar's test was used to compare categorical variables before and during confinement. Multiple linear regression was used to determine factors associated with stress during confinement. Logistic regression was used to assess the association between dietary habits and weight change (categorical outcomes) with stress scores and other factors. Models were adjusted for potential confounding variables; variables were included in the model if they had a bivariate correlation of $p \leq 0.2$ with the outcome. The number of events in the outcome was taken into consideration to limit the number of independent variables included in the logistic regression models [30]. All dietary habit models were adjusted for the dietary habit before confinement and fasting during Ramadan. During Ramadan, Muslims abstain from food and drink between dawn and dusk, with the exception of children under puberty, pregnant and lactating women, the sick and debilitated, and those who are traveling. All data were analyzed using Stata software version SE/14.2 for Mac (Stata Corp., College Station, TX: StataCorp LP) and the statistical significance was set to $p < 0.05$.

3. Results

Overall, $n = 503$ participants responded to the survey from 16 May to 28 June 2020. A total of 119 participants ($n = 119$) were excluded as they were not living in Saudi Arabia at the time and we wanted to minimize variability that could be attributed to differences in countries' restrictions. Participants <19 years old (n = 5), pregnant women (n = 8), breastfeeding women (n = 19), and those who were not under home confinement (n = 37) were excluded. After removing duplicates in Stata (n = 61), the final sample for analysis was n = 254 and their responses were received between 16 May and 12 June 2020.

Participants' characteristics are shown in Table 1. Mean ± SD age was 31.7 ± 8.8 years. The majority of the participants were females (87%), had a bachelor's degree or higher (91%), and almost half were married (48%). Moreover, most respondents were Saudis (88%), over half were from the Western region (67%), and 45% had children. The mean ± SD BMI was 26.2 ± 6.5 kg/m^2, and 49% of the participants were classified as overweight or obese. About 60% (n = 153) of the participants reported a change in their weight during confinement; 30.7% reported losing weight, while 29.5% reported gaining weight.

Table 1. General characteristics of the study participants.

Parameter	Total (n = 254)
Age, years	31.7 ± 8.8
Females	199 (78.4)
Marital status	
Single	121 (47.6)
Married	123 (48.4)
Separated/divorced	10 (3.9)
Lockdown > 40 days	201 (79.1)
Living region	
Western Region	171 (67.3)
Central Region	49 (19.3)
Eastern Region	22 (8.7)
Others	12 (4.7)
Education (BSc. Degree or higher)	231 (91.0)
Employment status	
Government	77 (30.3)
Private sector	40 (15.8)
Student	65 (25.6)
Unemployed	35 (13.8)
Other	37 (14.6)
Has children	114 (44.9)
Nationality	
Saudi	224 (88.2)
Current smokers	34 (13.4)
BMI, kg/m^2	26.2 ± 6.5
Underweight, BMI < 18.5	20 (7.9)
Healthy weight, 18.5 ≤ BMI < 25	109 (42.9)
Overweight obesity, BMI ≥ 25	125 (49.2)
Weight change during confinement	
Increased	75 (29.5)
Decreased	78 (30.7)
Same/do not know	101 (39.8)

Data are presented as n (%) except for age and BMI as mean ± SD.

3.1. Changes in Stress and Lifestyle Factors during Home Confinement

Stress components before and during home confinement are presented in Figure 1. Comparisons of stress scores, dietary habits, and other lifestyle factors before and during home confinement are shown in Table 2. Mean ± SD stress scores (depression subscale) during home confinement were higher than before home confinement (12.8 ± 7.5 vs. 10.7 ± 7.3, $p < 0.0001$). Similarly, the mean ± SD PSQI scores were higher during home confinement compared to before, indicating poorer sleep quality (7.3 ± 3.5 vs. 6.3 ± 3.3, $p < 0.0001$). For physical activity, the median number of days spent in walking for at least 10 min was reduced from 2 days before confinement to 1 day per week. However, no significant differences were observed in days spent engaging in moderate or vigorous activities before and during home confinement. Screen time was significantly different during home confinement compared to before; 58% reported spending more than 6 h on their phones, computers, TV, etc. during home confinement compared to 24% before confinement ($p < 0.001$).

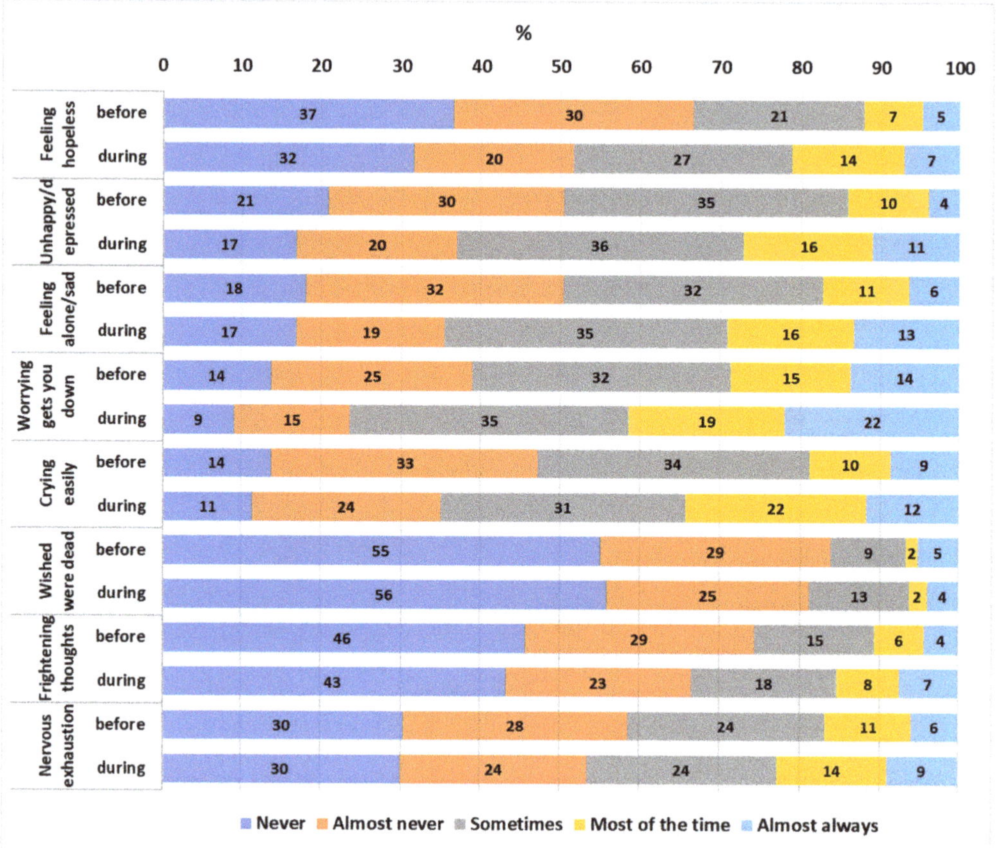

Figure 1. Percentages of stress responses before and during home confinement. All components were statistically significant ($p < 0.001$) before and during confinement, except for "wished were dead"), using Wilcoxon signed-rank test.

Some dietary habits have significantly changed during home confinement compared to before home confinement (Table 2). The proportions of adults consuming ≥three meals compared to one–two main meals decreased during home confinement, whereas consuming ≥two snacks per day increased during home confinement. Moreover, participants were less likely to consume outside food (restaurants, take-away, or delivered food) or eat at the same time during home confinement compared to before. There was no significant change in supplement use during home confinement compared to before confinement.

Table 2. Stress and lifestyle factors before and during home confinement.

Parameter	Before	During	*p*-Value
Stress score	10.7 ± 7.3	12.8 ± 7.5	<0.0001
PSQI score	6.29 ± 3.27	7.28 ± 3.50	<0.0001
Physical activity			
Days of vigorous activity	0 (0, 2)	0 (0, 2)	0.703
Days of moderate activity	0 (0, 2)	0 (0, 2)	0.506
Days of walking for at least 10 min	2 (0, 4)	1 (0, 3)	<0.001
Screen time			
Never	4 (1.6)	1 (0.4)	<0.001
1–3 h	84 (33.1)	28 (11.0)	
4–6 h	105 (41.3)	77 (30.3)	
>6 h	61 (24.0)	148 (58.3)	
Number of snacks			
<2 snacks	132 (52.0)	116 (45.7)	0.048
≥2 snacks	122 (48.0)	138 (54.3)	
Eating at the same time			<0.001
Always/most of the time	173 (68.1)	139 (54.7)	
Sometimes	61 (24.0)	81 (31.9)	
Never/almost never	20 (7.9)	34 (13.4)	
Supplement use			
Yes	139 (54.7)	144 (56.7)	0.424
No	115 (45.3)	110 (43.3)	
Main meals			0.012
1–2 meals	128 (50.4)	152 (59.8)	
≥3 meals	126 (49.6)	102 (40.2)	
Eating outside foods			<0.001
Never/less than once a week	95 (37.4)	217 (85.4)	
1–2 per week	111 (43.7)	26 (10.2)	
≥3 per week	48 (18.9)	11 (4.3)	

Data are presented as mean ± SD (for stress) and median (25, 75th percentile) for physical activity. McNemar's test was used to compare proportions. Paired *t*-test was used to compares stress and PSQI scores before and during confinement, while Wilcoxon signed-rank test was used for physical activity.

3.2. Predictors of Stress Scores during Confinement

Factors associated with total stress scores during confinement are shown in Table 3. Bivariate analysis showed that higher stress scores during confinement were associated with higher stress scores before confinement, poorer sleep quality, and weight change. However, lower stress scores during confinement were reported by those who spent more days walking and those who have children compared to those who do not have children. In multiple linear regression, only variables that have a *p*-value < 0.2 in the bivariate analyses were included in the model. Sleep quality during confinement, days spent in walking during confinement, and stress scores before confinement remained significantly associated with stress scores during home confinement.

3.3. Stress Scores and Eating Habits during Confinement

The association between stress scores during home confinement and eating habits is presented in Table 4. Higher stress scores during home confinement were associated with never/almost never eating food at the same time during confinement, as well as (sometimes) compared to eating food at the same time always or most of the time, adjusting for eating time variable before confinement. Stress scores during confinement remained significantly associated with eating at the same time after adjusting for smoking, sleep quality during confinement, currently fasting during Ramadan, and eating at the same time variable before confinement. Those who never eat at the same time during confinement were more likely to be smokers [OR (95% CI): 5.16 (1.66, 16.00)] and had poorer sleep quality [OR (95% CI): 1.16 (1.02, 1.33)] compared to those who always eat at the same time.

Table 3. Linear regression of stress scores during home confinement with participants characteristics and lifestyle factors.

Parameter	Bivariate B (95% CI)	Multiple B (95% CI)
Age, year	−0.096 (−0.20, 0.01)	−0.03 (−0.12, 0.05)
Sex		---
Female	0	
Male	−1.24 (−3.51, 1.02)	
Married (yes vs. no)	−1.04 (−2.91, 0.82)	---
Has children (yes vs. no)	−1.97 (−3.83, −0.10) *	0.22 (−1.30, 1.74)
Current smoker (yes vs. no)	0.31 (−2.43, 3.06)	---
Region		
Western region	0	0
Central region	−0.49 (−2.88, 1.91)	−0.28 (−1.86, 1.31)
Eastern region	−0.57 (−3.92, 2.78)	−0.37 (−2.55, 1.81)
Others	5.00 (0.58, 9.41) *	−0.14 (−3.05, 2.78)
Education (BSc. or higher)	0.64 (−2.61, 3.90)	---
Employment status		---
Government	0	
Private	−0.36 (−3.28, 2.55)	
Student	−0.02 (−2.54, 2.50)	
Others	−0.55 (−3.00, 1.90)	
BMI	0.00 (−0.14, 0.15)	---
Weight change (yes vs. no)	1.93 (0.04, 3.82) *	0.91 (−0.37, 2.19)
Stress scores before confinements	0.74 (0.65, 0.83) **	0.67 (0.58, 0.76) **
PSQI during confinement	0.93 (0.69, 1.17) **	0.53 (0.34, 0.71) **
Days of walking for at least 10 min	−0.44 (−0.84, −0.03) *	−0.42 (−0.69, −0.15) **

* $p < 0.05$; ** $p < 0.01$.

Table 4. Association of stress scores with eating habits during home confinement.

Outcome	Model 1 OR (95% CI)	*p*	Model 2 OR (95% CI)	*p*
(A) Eating at the same time				
Always/most of the time	1 (reference group)		1 (reference group)	
Sometimes	1.07 (1.02, 1.11)	0.002	1.07 (1.02, 1.12)	0.005
Never/almost never	1.11 (1.05, 1.18)	<0.001	1.09 (1.02, 1.16)	0.008
(B) Number of main meals				
1–2 meals	1 (reference group)		1 (reference group)	
≥3 meals	0.94 (0.91, 0.98)	0.002	0.95 (0.91, 0.99)	0.013
(C) Eating outside food				
Never/less than once a week	1 (reference group)		1 (reference group)	
1–2 times per week	0.96 (0.90, 1.02)	0.201	0.95 (0.89, 1.02)	0.148
≥3 times per week	1.12 (1.03, 1.22)	0.009	1.08 (0.99, 1.19)	0.076

Logistic regression (n = 254) was used for all outcomes: A, B, and C. Model 1 is adjusted for dietary habits before home confinement. Model 2 is adjusted for (A) smoking, sleep quality, fasting during Ramadan, and eating time before home confinement; (B) age, smoking, sleep quality during confinement, currently fasting during Ramadan, and the number of meals before confinement; (C) sex, BMI, marital status, sleep quality during confinement, currently fasting during Ramadan, smoking, and eating outside before confinement. An OR of 1.11 indicates that the odds of never/almost never eating at the same time during home confinement are 1.11 times higher with 1-unit increase in stress scores after holding smoking, sleep quality, fasting during Ramadan, and eating time before home confinement at fixed values.

Participants who reported consuming ≥three meals during confinement had lower stress scores during confinement than those who reported consuming one–two meals, adjusting for the number of meals before confinement. This relationship remained significant after adjusting for age, smoking, sleep quality during confinement, currently fasting during Ramadan, and the number of meals before confinement. Consuming outside food (take away, dine out, or delivery) ≥three times per week during confinement was associated with higher stress scores compared to those who reported never consume outside food or less than once a week. This, however, was not significant when adjusting for sex, BMI, marital status, sleep quality during confinement, currently fasting during Ramadan, and smoking.

Gaining or losing weight during confinement was not associated with stress scores during confinement in the bivariate analyses. The odds of gaining weight during confinement was associated with fewer days spent walking for at least 10 min, being married, poorer sleep quality, and consuming ≥three meals compared to one–two meals during confinement (Table 5). Higher odds of losing weight were associated with more days spent doing moderate activity, whereas losing weight was less likely with consuming ≥three meals compared to one–two meals during confinement.

Table 5. Predictors of weight gain or loss during home confinement.

Variables	OR (95% CI)	*p*
Weight Gain (reference: no weight gain)		
BMI, kg/m^2	1.03 (0.98, 1.08)	0.211
Sex (male vs. females)	1.54 (0.77, 3.08)	0.224
Married (yes vs. no)	2.40 (1.31, 4.42)	0.005
PSQI during confinement	1.10 (1.01, 1.20)	0.025
Days of walking for at least 10 min	0.82 (0.70, 0.95)	0.007
Number of snacks (≥2 snacks vs. <2 snacks)	1.83 (0.98, 3.42)	0.056
Number of main meals (≥3 meals vs. 1–2 meals)	1.76 (0.95, 3.26)	0.07
Weight loss (Reference: no weight loss)		
Age, years	0.97 (0.93, 1.02)	0.228
Married (yes vs. no)	0.93 (0.46, 1.87)	0.833
Number of main meals (≥3 meals vs. 1–2 meals)	0.36 (0.19, 0.67)	0.001
Current smoker (yes vs. no)	0.49 (0.19, 1.29)	0.149
Days of moderate Activity	1.25 (1.07, 1.46)	0.004

Multinomial logistic regression (n = 254) was used.

4. Discussion

Restrictions imposed to control the COVID-19 pandemic have been reported as negatively affecting lifestyle habits in many countries. These restrictions, however, differed between countries and researchers are still exploring the different impacts to tackle adverse effects on the individual's well-being in future lockdowns. In this sample of Saudi adults, the majority had been under home confinement for >40 days and reported poorer sleep quality, higher stress scores, and altered dietary habits compared with before confinement. During home confinement, more people consumed only one–two meals, never consumed outside food, and consumed ≥two snacks per day. Higher stress scores during home confinement were associated with never consuming food at the same time, whereas lower stress scores were associated with consuming ≥three meals compared to one–two meals. Weight gain was more likely among those who were married and had poorer sleep quality, but less likely with more days spent in walking.

Approximately 68% of study participants aged 35 years or younger and the majority of our sample were from the Western region that faced more strict regulations, including a 24-h curfew, compared to the rest of the country [2,31], making it an extreme case. However, stress scores during home confinement did not differ between individuals living in the Western region compared to the remainder who did not live in the Western region. The "other regions" category mainly represented participants from the Southern region, which showed higher stress scores during confinement compared to other regions. This relationship did not hold when adjusting for stress scores before confinement, indicating that the stress level before confinement drove the relationship.

We found higher stress scores during home confinement than before confinement, which is in line with previous reports in the Saudi population and other countries [6,13,16,18,19]. High stress scores before confinement predicted high stress scores during confinement, an observation that is expected but not previously reported in the context of COVID-19. As expected, stress during confinement was higher with poorer sleep quality, independent of stress level before confinement. Similar findings were reported during COVID-19 lockdown [19,32–34]. Sleep and stress can be bidirectional, with stressors affecting sleep quality and vice versa. Interestingly, those who had children reported lower stress scores during confinement than those who did not have children in bivariate analyses. A study of 4920 healthcare providers in Saudi Arabia during the pandemic reported lower anxiety levels among healthcare providers who have children, but not when adjusting for potential confounders [35]. A study on 838 U.S. adults during the last week of April 2020 reported that mean stress scores, as assessed using the Perceived Stress Scale, differed by ethnic group and were higher among those with part-time jobs, those who were unmarried, and those who were under 35.

Food is a major factor in determining health, and the stress caused by the pandemic influenced how people eat. Several studies have reported changes in dietary habits, but very few have used measures of stress to explore its effect on diets. In our study, dietary habits have changed during the COVID-19 pandemic, including more snacking, variation in the time of eating, fewer main meals, and reduction in consuming foods from outside the home. Some of these changes were associated with higher stress scores during confinement. The frequency of eating food from outside the home was reduced during confinement, similar to other studies in Saudi Arabia [36], which was to be expected given the uncertainty surrounding the safety of outside food in the pandemic and whether the virus could be transmitted through food. Interestingly, those who consumed outside food ≥3 times per week during the confinement tended to have higher stress scores, adjusting for consuming outside food before the pandemic. This relationship was not significant when adjusting for sex, BMI, marital status, sleep quality during confinement, currently fasting during Ramadan, and smoking, but worth further investigation.

In addition to the variation in the time of eating during confinement, higher stress scores during home confinement seem to disturb the day-to-day timing of consuming food. Those who reported never consuming the meals at the same time (eating misalignment) had higher stress scores even after adjusting for smoking, sleep quality, fasting during Ramadan, and dietary habits before home confinement. Stress generally has been reported to affect the quantity of food consumed and food choices [12,37,38], but its impact on eating misalignment is not well-established. The consumption of food at times that contradict our circadian system may have implications for weight status and metabolic health [39,40], and disrupt the synchronization of peripheral clock tissues, such as the liver, leading to metabolic dysfunction [41]. Studies have suggested that shift workers, who often experience circadian misalignment due to irregular sleep–wake and eating patterns, are at an increased risk of developing metabolic syndrome and cardiovascular disease [42]. We believe that no other studies have explored the relationship between stress and consuming meals at set times during the pandemic and this need to be confirmed in other populations, taking into consideration multiple factors [43].

During home confinement, fewer people consumed ≥three meals but more consumed ≥two snacks per day, which could be attributed to skipping meals, as previously reported [6]. However, conflicting findings have been reported on skipping meals. A study in the United Arab Emirates reported lower rates of skipping meals during lockdown and that the lack of time was the main reason for skipping meals before lockdown [13]. A lack of appetite may be the main contributor to skipping meals during the pandemic [6,13]. After adjusting for confounders, mean stress scores remained lower among those who consumed ≥three meals than one–two meals. Similarly, stress in US adults was associated with reducing meals more during the pandemic compared to before [18]. Increased control of over-eating during the lockdown was negatively associated with anxious feelings, but not with depressed mood in Italian adults; questions on anxiety and depression were extrapolated from the Hamilton Anxiety Rating Scale and the Hamilton Depression Scale [44]. Nearly twice the increase in reported anxiety among adults from several countries were observed in association with a perceived change toward unhealthy eating in bivariate analyses [17].

A study on $n = 1047$ adults from several countries reported that unhealthy eating habits and poor sleep were linked to low mental wellness, life dissatisfaction, and high levels of depressive symptoms using multiple questionnaires [34]; it is important to note that only bivariate analyses were conducted in this study without adjustments for any confounders. Binge-drinking alcohol, a negative dietary behavior, was higher in Indian adults ($n = 354$) who were more depressed, as assessed using the Depression Anxiety Stress Scale, during the 2020 COVID-19 lockdown [45]; this study also did not adjust for confounders and used only binary logistic regression. Adults from the US who reported changes in their eating practices (either more or less than before the pandemic) and those reporting that their diet had deteriorated had significantly higher stress scores after controlling for demographic characteristics and BMI [18].

The quality of the diet could be also affected by the stress of the COVID-19 lockdown, but our study focused on dietary habits rather than dietary content. A study on Ecuadorian adults assessing dietary quality during COVID-19 reported higher levels of stress (measured by perceived stress score) with lower dietary quality (measured by the global diet index) [46]. An Italian study concluded that 52.9% were eating more during the lockdown and that there had been an increase in comfort food consumption, including chocolate, ice cream, desserts, and salty snacks [47]. However, not all studies reported low dietary quality during COVID-19 restrictions. A study on 10 Arab countries during the pandemic reported an increase in fruit consumption, unprocessed meats, and lower sugary drinks' consumption, reflecting positive changes [48]. A similar study in Italy reported a reduction in processed meat and a higher consumption of vegetables, but increased sweets' consumption [49]. A study conducted in early 2021 on adults in India also reported positive changes in eating habits, such as eating more nutrient-dense food and eating more home-cooked meals, but there was an increase in binge eating, snacking, and larger portion sizes at meals [15]. In US adults, similar findings were reported in terms of a decreased intake of fast food, an increased intake of fruit, and an increased intake of sweets [17]. Dietary factors may help to reduce stress via improving the body's ability to regulate stress hormones. Overweight and obese women who followed dietary guidelines for 8 weeks had significantly lower levels of stress than those following the traditional American diet; researchers attributed these findings to the combination of higher vegetable intake and a lower sodium intake [50]. Additionally, increasing dietary carbohydrate intake as part of a healthy whole food diet intervention in these women dampened the stress-related changes in salivary cortisol and cortisol responsiveness [51]. Authors suggest that this may be due to the fact that carbohydrate intake helps to stabilize blood sugar levels, which can reduce the body's need to produce stress hormones.

The Saudi population has been witnessing an increase in poor diets, stress, and obesity over the past few decades, which are associated with adverse health outcomes [52,53]. As a result of the COVID-19 pandemic, it has been proposed that stress and poor diet were exacerbated, creating additional barriers to healthy behaviors. The combined rates of overweight (25%) and obesity (24%) were slightly lower than the national estimates in 2019, when overweight and obesity were 38% and 20%, respectively [53]. Stress is proposed to contribute to overweight and obesity through cognitive processes, behavioral effects, physiological changes in the hypothalamic–pituitary–adrenal axis, and promotion of the production of biochemical hormones and peptides such as leptin, ghrelin, and neuropeptide Y [37]; behavioral effects include overeating, reduced sleep quality, and increased physical inactivity.

Restrictions during COVID-19 in Saudi Arabia were at their highest during the Ramadan and Eid-ul-Fitr holidays, which coincided with our study. This could have affected eating and sleeping habits; however, we controlled for fasting during Ramadan in our dietary models. None of the other studies considered this in their models. Ramadan is a spiritual month in the Arabic lunar calendar practiced by Muslims worldwide and has important social aspects. In 2020, Ramadan was between 23 April to 23 May 2020, and our data collection started on 16 May. In the last ten days of Ramadan, corresponding to 14 May 2020, people were allowed to move freely from 9 A.M. to 5 P.M., in all cities and regions, except for Makkah city where curfew continued [54]. A 24-h curfew was imposed for the first five days of the Eid Al-Fitr holiday (23 May to 27 May 2020) in all cities [55,56]. Intermittent fasting has been shown to affect the circadian system and affect sleep, although subsequent research that took into consideration lifestyle choices and sleep/wake cycles found no differences in biological clock markers, daytime tiredness, or sleep characteristics [57]. The vast cultural and geographic diversity across groups makes it difficult to draw broad conclusions about changes in body weight and diet composition during Ramadan. However, numerous studies show that people do not tend to reduce the number of calories they consume daily during the fasting period [58]. Fasting during Ramadan has been associated with less anxiety in Iranians during the COVID-19 wave in 2021 [59].

Approximately 30% of study participants reported weight gain, which is comparable to the 30.3% in the Middle East and North Africa region [6], and 27.5% worldwide [17]. Estimates in weight gain due to COVID-19-related lockdown varied between populations. For example, weight gain during the lockdown period was 18% in Saudi women [16], 19.5–48.6% in Italy [47,60], 31% in the United Arab Emirates [13], and 22% in US adults [21]. The odds of gaining weight during home confinement were higher among married participants, those with poorer sleep quality, and those with fewer days of walking. Sleep duration has been associated with BMI variation [61]. A study on 173 adults from the US reported that weight gain was associated with shorter sleep duration and less physical activity, but reported null finding with the Perceived Stress Scale [21]. We did not observe sex differences in weight change and most other parameters, which could be because over 70% of the sample were females. Adults over the age of 35 years and males appear to be more likely to gain weight during lockdown [13,48]. However, a study reviewing hospital records from three different regions in Saudi Arabia before and after 2020 reported that females and younger adults have higher odds of gaining $\geq$5% of their pre-2020 weight; pregnant women do not appear to be excluded, which might have affected these results [62]. Similarly, a study on Danish adults reported that women were more likely to gain weight during the pandemic [14].

Although higher odds of gaining weight were observed with more snacking and a higher number of meals consumed, this was not significant ($p = 0.06$ and 0.07). Young adults who reported overeating to cope with the pandemic gained more weight from baseline to follow-up; however, unhealthy food intake to cope with the pandemic was not associated with weight change [63]. The magnitude of weight gain seems to be a factor, as US adults who gained 5–10 pounds reported more snacking after dinner and eating more in response to stress during the pandemic, but not the other groups of weight change [21]. The odds of losing weight were lower with consuming $\geq$three meals compared to one–two meals, and higher with increasing days spent in moderate activity. Neither stress scores during confinement nor fasting during Ramadan were associated with odds of weight increase or weight decrease in our sample.

Higher stress scores were associated with higher odds of weight change in our study, but not after adjusting for age, region, sleep quality, physical activity during confinement, and stress score before confinement. A prospective study in Saudi women ($n = 297$) reported no association between stress scores, assessed using the Perceived Stress Scale, and weight change as an outcome in the adjusted models, although baseline stress might be associated with weight gain [16]. A study on adults reported that anxiety scores were higher among those who reported weight change during the lockdown compared to those who stayed the same [17]. Many factors could cause weight gain and loss and conflicting findings between studies could be attributed to not controlling for these factors that could be driving the relationship, and potentially the different tools used to assess stress, anxiety, or depression. A study on n = 521 adolescents and young adults in Peru reported that lockdown-related changes in sleep duration, screen exposure time, and sedentarism explained only 6% of the variation in BMI [61]. The weight before confinement was not available in our study and can be a confounding factor in predicting weight change [63]. Other considerations in drawing conclusions on mental health relationships with dietary patterns or weight change are the differences in COVID-19 restrictions as well as timing between nations.

Our study has some strengths. We targeted adults under extremely restrictive conditions during the Spring of 2020. Standardized questionnaires were used to assess sleep, stress, and physical activity, which allowed comparisons with other studies. The questionnaires were administered in English and Arabic to facilitate reachability. Data were collected on before and during home confinement, which provided comparisons and allowed for statistical adjustments of the factors before confinement.

There are some limitations that need to be acknowledged. The cross-sectional nature of the study limits the ability to establish causal relationships between the study's variables. The reliance on self-reported data may have resulted in some recall, social desirability, and/or recency biases as recent behaviors and emotions would be easier to recall than those that took place in the past. In addition, the sample was collected using an online platform that needed an internet connection and a certain level of literacy to answer the survey, limiting the generalizability of the findings to all adults in Saudi Arabia. For example, people who are less literate or who do not have access to the internet may have been less likely to participate in the study. This could have biased the results towards people who are more educated and have more access to resources. However, this would have a minimal impact given that 98% of individuals in Saudi Arabia had access to the internet in 2020 [64]. Most respondents completed the survey during Ramadan, when all physically capable Muslims are required to fast. Even though it was made explicit in the questionnaire that it was only about the lockdown, it is possible that this had an impact on how participants responded; fasting during Ramadan was adjusted for in the dietary habit models. Moreover, the length of the questionnaire may have introduced a bias towards participants who were more health conscious, as individuals who are highly attentive to their well-being or research-oriented might be more likely to participate and complete the lengthy questionnaire. This self-selection bias could influence the study's results through overrepresenting individuals who are already more conscious of their health, potentially affecting the generalizability of the findings to the broader population. Details on food items consumed were not available, which limited our ability to adjust for food quantity and energy intake in the weight change models. Specifics about the diet consumed could provide a closer look at the participants' diet quality and patterns that are factors with a bidirectional relationship with mental health.

5. Conclusions

This study explored the lifestyle changes and psychological impact of COVID-19, and the associations between stress, dietary habits, and weight change during the COVID-19 lockdown in Saudi Arabia. Our findings illustrate a significant impact of home confinement on mental health and on unhealthy lifestyle during the COVID-19 pandemic, as opposed to prior confinement. The stress level during home confinement was linked to changes in the timing and number of meals but not weight change. Studies need to consider potential social determinants and confounding factors when studying the impact of lockdowns on the individual's behavioral and physical changes before drawing conclusions on bivariate analyses. These multifaceted negative impacts emphasize how crucial it is for stakeholders and policymakers to think about creating and putting into practice interdisciplinary solutions to lessen the physical and psychological stress that such pandemics elicit.

Author Contributions: Conception and design, E.A.N. and A.M.A.; methodology, E.A.N.; data collection, E.A.N.; statistical analysis, A.M.A.; data interpretation, A.M.A., R.J.B., N.A.A., S.O.A. and E.A.N.; writing—original draft: A.M.A., R.J.B., N.A.A. and S.O.A.; writing—review and editing: A.M.A. and E.A.N. All authors have read and agreed to the published version of the manuscript.

Funding: This research received no external funding.

Institutional Review Board Statement: The study was conducted in accordance with the Declaration of Helsinki, and approved by the Ethics Committee of the Research and Studies Department, Directorate of Health Affairs—Taif, Saudi Arabia (HAP-02-T-067—Approval number: 357).

Informed Consent Statement: Informed consent was obtained from all subjects involved in the study.

Data Availability Statement: Data can be provided upon request from the corresponding author.

Conflicts of Interest: The authors declare no conflict of interest.

References

1. Director-General, W. WHO Director-General's Opening Remarks at the Media Briefing on COVID-19-11 March 2020. Available online: https://www.who.int/director-general/speeches/detail/who-director-general-s-opening-remarks-at-the-media-briefing-on-covid-19{-}{-}-11-march-2020 (accessed on 27 July 2022).
2. Khan, A.; Alsofayan, Y.; Alahmari, A.; Alowais, J.; Algwizani, A.; Alserehi, H.; Assiri, A.; Jokhdar, H. COVID-19 in Saudi Arabia: The National Health Response. *East. Mediterr. Health J.* **2021**, *27*, 1114–1124. [CrossRef]
3. Coronavirus Disease 2019 (COVID-19) Situation Report–51. Available online: https://www.who.int/publications/m/item/situation-report{-}{-}-51 (accessed on 15 April 2022).
4. Alhusseini, N.; Alammari, D.; Ramadan, M.; Ziadeh, N.; Zyadeh, Z.; Alshamrani, J.; Qasim, H.; Alamri, N.; Alqahtani, S.A. The Impact of COVID-19 Pandemic on Lifestyle among the Saudi Population. *J. Public Health Res.* **2022**, *11*, 227990362211231. [CrossRef]
5. Pellegrini, M.; Ponzo, V.; Rosato, R.; Scumaci, E.; Goitre, I.; Benso, A.; Belcastro, S.; Crespi, C.; Michieli, F.D.; Ghigo, E.; et al. Changes in Weight and Nutritional Habits in Adults with Obesity during the "Lockdown" Period Caused by the COVID-19 Virus Emergency. *Nutrients* **2020**, *12*, 2016. [CrossRef] [PubMed]
6. Cheikh Ismail, L.; Osaili, T.M.; Mohamad, M.N.; Al Marzouqi, A.; Jarrar, A.H.; Zampelas, A.; Habib-Mourad, C.; Omar Abu Jamous, D.; Ali, H.I.; Al Sabbah, H.; et al. Assessment of Eating Habits and Lifestyle during the Coronavirus 2019 Pandemic in the Middle East and North Africa Region: A Cross-Sectional Study. *Br. J. Nutr.* **2021**, *126*, 757–766. [CrossRef]
7. Alghamdi, B.S.; Alatawi, Y.; Alshehri, F.S.; Tayeb, H.O.; Tarazi, F.I. Relationship Between Public Mental Health and Immune Status During the COVID-19 Pandemic: Cross-Sectional Data from Saudi Arabia. *Risk Manag. Healthc. Policy* **2021**, *14*, 1439–1447. [CrossRef]
8. López-Moreno, M.; López, M.T.I.; Miguel, M.; Garcés-Rimón, M. Physical and Psychological Effects Related to Food Habits and Lifestyle Changes Derived from Covid-19 Home Confinement in the Spanish Population. *Nutrients* **2020**, *12*, 3445. [CrossRef] [PubMed]
9. Brooks, S.K.; Webster, R.K.; Smith, L.E.; Woodland, L.; Wessely, S.; Greenberg, N.; Rubin, G.J. The Psychological Impact of Quarantine and How to Reduce It: Rapid Review of the Evidence. *Lancet* **2020**, *395*, 912–920. [CrossRef] [PubMed]
10. Liu, Y.Z.; Wang, Y.X.; Jiang, C.L. Inflammation: The Common Pathway of Stress-Related Diseases. *Front. Hum. Neurosci.* **2017**, *11*, 316. [CrossRef]
11. Bacaro, V.; Ballesio, A.; Cerolini, S.; Vacca, M.; Poggiogalle, E.; Donini, L.M.; Lucidi, F.; Lombardo, C. Sleep Duration and Obesity in Adulthood: An Updated Systematic Review and Meta-Analysis. *Obes. Res. Clin. Pract.* **2020**, *14*, 301–309. [CrossRef]
12. Block, J.P.; He, Y.; Zaslavsky, A.M.; Ding, L.; Ayanian, J.Z. Psychosocial Stress and Change in Weight among US Adults. *Am. J. Epidemiol.* **2009**, *170*, 181–192. [CrossRef]
13. Ismail, L.C.; Osaili, T.M.; Mohamad, M.N.; Marzouqi, A.A.; Jarrar, A.H.; Jamous, D.O.A.; Magriplis, E.; Ali, H.I.; Sabbah, H.A.; Hasan, H.; et al. Eating Habits and Lifestyle during COVID-19 Lockdown in the United Arab Emirates: A Cross-Sectional Study. *Nutrients* **2020**, *12*, 3314. [CrossRef] [PubMed]
14. Giacalone, D.; Frøst, M.B.; Rodríguez-Pérez, C. Reported Changes in Dietary Habits During the COVID-19 Lockdown in the Danish Population: The Danish COVIDiet Study. *Front. Nutr.* **2020**, *7*, 592112. [CrossRef] [PubMed]
15. Madan, J.; Blonquist, T.; Rao, E.; Marwaha, A.; Mehra, J.; Bharti, R.; Sharma, N.; Samaddar, R.; Pandey, S.; Mah, E.; et al. Effect of COVID-19 Pandemic-Induced Dietary and Lifestyle Changes and Their Associations with Perceived Health Status and Self-Reported Body Weight Changes in India: A Cross-Sectional Survey. *Nutrients* **2021**, *13*, 3682. [CrossRef] [PubMed]
16. Al-Musharaf, S.; Aljuraiban, G.; Bogis, R.; Alnafisah, R.; Aldhwayan, M.; Tahrani, A. Lifestyle Changes Associated with COVID-19 Quarantine among Young Saudi Women: A Prospective Study. *PLoS ONE* **2021**, *16*, e0250625. [CrossRef] [PubMed]
17. Flanagan, E.W.; Beyl, R.A.; Fearnbach, S.N.; Altazan, A.D.; Martin, C.K.; Redman, L.M. The Impact of COVID-19 Stay-At-Home Orders on Health Behaviors in Adults. *Obesity* **2021**, *29*, 438. [CrossRef]

18. Khubchandani, J.; Kandiah, J.; Saiki, D. The COVID-19 Pandemic, Stress, and Eating Practices in the United States. *Eur. J. Investig. Health Psychol. Educ.* **2020**, *10*, 950–956. [CrossRef]
19. Franceschini, C.; Musetti, A.; Zenesini, C.; Palagini, L.; Scarpelli, S.; Quattropani, M.C.; Lenzo, V.; Freda, M.F.; Lemmo, D.; Vegni, E.; et al. Poor Sleep Quality and Its Consequences on Mental Health during the COVID-19 Lockdown in Italy. *Front. Psychol.* **2020**, *11*, 3072. [CrossRef]
20. Agurto, H.S.; Alcantara-Diaz, A.L.; Espinet-Coll, E.; Toro-Huamanchumo, C.J. Eating Habits, Lifestyle Behaviors and Stress during the COVID-19 Pandemic Quarantine among Peruvian Adults. *PeerJ* **2021**, *9*. [CrossRef]
21. Zachary, Z.; Brianna, F.; Brianna, L.; Garrett, P.; Jade, W.; Alyssa, D.; Mikayla, K. Self-Quarantine and Weight Gain Related Risk Factors during the COVID-19 Pandemic. *Obes. Res. Clin. Pract.* **2020**, *14*, 210–216. [CrossRef]
22. Khan, A.; Alahmari, A.; Almuzaini, Y.; Alturki, N.; Aburas, A.; Alamri, F.A.; Albagami, M.; Alzaid, M.; Alharbi, T.; Alomar, R.; et al. The Role of Digital Technology in Responding to COVID-19 Pandemic: Saudi Arabia's Experience. *Risk Manag. Heal. Policy* **2021**, *14*, 3923. [CrossRef]
23. Sedov, I.D.; Cameron, E.E.; Madigan, S.; Tomfohr-Madsen, L.M. Sleep Quality during Pregnancy: A Meta-Analysis. *Sleep Med. Rev.* **2018**, *38*, 168–176. [CrossRef] [PubMed]
24. Demirci, J.R.; Braxter, B.J.; Chasens, E.R. Breastfeeding and Short Sleep Duration in Mothers and 6–11-Month-Old Infants. *Infant. Behav. Dev.* **2012**, *35*, 884–886. [CrossRef] [PubMed]
25. Guardino, C.M.; Dunkel Schetter, C. Coping during Pregnancy: A Systematic Review and Recommendations. *Health Psychol. Rev.* **2014**, *8*, 70–94. [CrossRef] [PubMed]
26. IPAQ Group Guidelines for the Data Processing and Analysis of the International Physical Activity Questionnaire. Available online: https://sites.google.com/view/ipaq/score (accessed on 1 August 2017).
27. World Health Organization STEPS Manual: Section 5: Collecting Step 2 Data: Physical Measurements. Available online: https://cdn.who.int/media/docs/default-source/ncds/ncd-surveillance/steps/part3-section5.pdf?sfvrsn=a46653c7_2 (accessed on 19 April 2018).
28. Buysse, D.J.; Reynolds, C.F.; Charles, F.; Monk, T.H.; Berman, S.R.; Kupfer, D.J. Pittsburgh Sleep Quality Index (PSQI). *Psychiatry Res.* **1989**, *28*, 193–213. [CrossRef]
29. Carlson, L.E.; Cherian Thomas, B. Development of the Calgary Symptoms of Stress Inventory (C-SOSI). *Int. J. Behav. Med. Copyr. C* **2007**, *14*, 249–256. [CrossRef]
30. Peduzzi, P.; Concato, J.; Kemper, E.; Holford, T.R.; Feinstein, A.R. A Simulation Study of the Number of Events per Variable in Logistic Regression Analysis. *J. Clin. Epidemiol.* **1996**, *49*, 1373–1379. [CrossRef]
31. Alfawaz, H.; Amer, O.E.; Aljumah, A.A.; Aldisi, D.A.; Enani, M.A.; Aljohani, N.J.; Alotaibi, N.H.; Alshingetti, N.; Alomar, S.Y.; Khattak, M.N.K.; et al. Effects of Home Quarantine during COVID-19 Lockdown on Physical Activity and Dietary Habits of Adults in Saudi Arabia. *Sci. Rep.* **2021**, *11*, 5904. [CrossRef]
32. Cellini, N.; Canale, N.; Mioni, G.; Costa, S. Changes in Sleep Pattern, Sense of Time and Digital Media Use during COVID-19 Lockdown in Italy. *J. Sleep Res.* **2020**, *29*, e13074. [CrossRef]
33. Alqahtani, S.S.; Banji, D.; Banji, O.J.F.F. A Survey Assessing Sleep Efficiency among Saudis during COVID-19 Home Confinement Using the Pittsburgh Sleep Quality Index: A Call for Health Education. *Saudi. Pharm. J.* **2021**, *29*, 692–698. [CrossRef]
34. Ammar, A.; Trabelsi, K.; Brach, M.; Chtourou, H.; Boukhris, O.; Masmoudi, L.; Bouaziz, B.; Bentlage, E.; How, D.; Ahmed, M.; et al. Effects of Home Confinement on Mental Health and Lifestyle Behaviours during the COVID-19 Outbreak: Insight from the ECLB-COVID19 Multicenter Study. *Biol. Sport* **2020**, *38*, 9–21. [CrossRef]
35. Alenazi, T.H.; BinDhim, N.F.; Alenazi, M.H.; Tamim, H.; Almagrabi, R.S.; Aljohani, S.M.; Basyouni, M.H.; Almubark, R.A.; Althumiri, N.A.; Alqahtani, S.A. Prevalence and Predictors of Anxiety among Healthcare Workers in Saudi Arabia during the COVID-19 Pandemic. *J. Infect. Public Health* **2020**, *13*, 1645–1651. [CrossRef] [PubMed]
36. Alhusseini, N.; Alqahtani, A. COVID-19 Pandemic's Impact on Eating Habits in Saudi Arabia. *J. Public Health Res.* **2020**, *9*, 354–360. [CrossRef] [PubMed]
37. Tomiyama, A.J. Stress and Obesity. *Annu. Rev. Psychol* **2019**, *70*, 703–718. [CrossRef] [PubMed]
38. Costarelli, V.; Michou, M. Perceived Stress Negatively Affects Diet Quality and Life Satisfaction during the COVID-19 Lockdown Period, in Greece. *Nutr. Food Sci.* 2023; *ahead-of-press*. [CrossRef]
39. Chaput, J.P.; McHill, A.W.; Cox, R.C.; Broussard, J.L.; Dutil, C.; da Costa, B.G.G.; Sampasa-Kanyinga, H.; Wright, K.P. The Role of Insufficient Sleep and Circadian Misalignment in Obesity. *Nat. Rev. Endocrinol.* **2022**, *19*, 82–97. [CrossRef]
40. BaHammam, A.S.; Pirzada, A. Timing Matters: The Interplay between Early Mealtime, Circadian Rhythms, Gene Expression, Circadian Hormones, and Metabolism—A Narrative Review. *Clocks Sleep* **2023**, *5*, 507–535. [CrossRef]
41. Binks, H.; Vincent, G.E.; Gupta, C.; Irwin, C.; Khalesi, S. Effects of Diet on Sleep: A Narrative Review. *Nutrients* **2020**, *12*, 936. [CrossRef]
42. Schroor, M.M.; Sennels, H.P.; Fahrenkrug, J.; Jørgensen, H.L.; Plat, J.; Mensink, R.P. Diurnal Variation of Markers for Cholesterol Synthesis, Cholesterol Absorption, and Bile Acid Synthesis: A Systematic Review and the Bispebjerg Study of Diurnal Variations. *Nutrients* **2019**, *11*, 1439. [CrossRef]
43. Dashti, H.S.; Scheer, F.A.J.L.; Saxena, R.; Garaulet, M. Timing of Food Intake: Identifying Contributing Factors to Design Effective Interventions. *Adv. Nutr.* **2019**, *10*, 606–620. [CrossRef]

44. Renzo, L.D.; Gualtieri, P.; Cinelli, G.; Bigioni, G.; Soldati, L.; Attinà, A.; Bianco, F.F.; Caparello, G.; Camodeca, V.; Carrano, E.; et al. Psychological Aspects and Eating Habits during COVID-19 Home Confinement: Results of EHLC-COVID-19 Italian Online Survey. *Nutrients* **2020**, *12*, 2152. [CrossRef]
45. Verma, S.; Mishra, A. Depression, Anxiety, and Stress and Socio-Demographic Correlates among General Indian Public during COVID-19. *Int. J. Soc. Psychiatry* **2020**, *66*, 756–762. [CrossRef]
46. Abril-Ulloa, V.; Santos, S.P.L.d.; Morejón-Terán, Y.A.; Carpio-Arias, T.V.; Espinoza-Fajardo, A.C.; Vinueza-Veloz, M.F. Stress and Diet Quality Among Ecuadorian Adults During the COVID-19 Pandemic. A Cross-Sectional Study. *Front. Nutr.* **2022**, *9*, 924076. [CrossRef]
47. Scarmozzino, F.; Visioli, F. COVID-19 and the Subsequent Lockdown Modified Dietary Habits of Almost Half the Population in an Italian Sample. *Foods* **2020**, *9*, 675. [CrossRef] [PubMed]
48. Tayyem, R.; Ibrahim, M.O.; Mortada, H.; Alkhalaf, M.; Bookari, K.; Al Sabbah, H.; Qasrawi, R.; Kamel, I.; Dashti, S.; Allehdan, S.; et al. Sex Disparities in Food Consumption Patterns, Dietary Diversity and Determinants of Self-Reported Body Weight Changes before and amid the COVID-19 Pandemic in 10 Arab Countries. *Front. Public Health* **2022**, *10*, 3817. [CrossRef] [PubMed]
49. Ferrante, G.; Camussi, E.; Piccinelli, C.; Senore, C.; Armaroli, P.; Ortale, A.; Garena, F.; Giordano, L. Did Social Isolation during the SARS-CoV-2 Epidemic Have an Impact on the Lifestyles of Citizens? *Epidemiol. Prev.* **2020**, *44*, 353–362. [CrossRef] [PubMed]
50. Soltani, H.; Keim, N.L.; Laugero, K.D. Diet Quality for Sodium and Vegetables Mediate Effects of Whole Food Diets on 8-Week Changes in Stress Load. *Nutrients* **2018**, *10*, 1606. [CrossRef] [PubMed]
51. Soltani, H.; Keim, N.L.; Laugero, K.D. Increasing Dietary Carbohydrate as Part of a Healthy Whole Food Diet Intervention Dampens Eight Week Changes in Salivary Cortisol and Cortisol Responsiveness. *Nutrients* **2019**, *11*, 2563. [CrossRef] [PubMed]
52. DeNicola, E.; Aburizaiza, O.S.; Siddique, A.; Khwaja, H.; Carpenter, D.O. Obesity and Public Health in the Kingdom of Saudi Arabia. *Rev Env. Health* **2015**, *30*, 191–205. [CrossRef]
53. World Health Survey-Saudi Arabia. Available online: https://www.moh.gov.sa/en/Ministry/Statistics/Population-Health-Indicators/Documents/World-Health-Survey-Saudi-Arabia.pdf (accessed on 5 April 2023).
54. MOI Agency for Security Capabilities All Precautionary Measures Continued till the End of Ramadan, a Complete Curfew During the Period during 30/09-4/10/1441 Ah. Available online: https://www.moi.gov.sa/wps/portal/Home/sectors/mafpasd/contents/!ut/p/z1/jZHNTsMwEIRfJZccK49_QszRAhG7REItCg2-RFaJWiPi0BI18PYkoVfS-rDyjr_ZXa2JJSWxwZ38znW-De5jyF_tTUXpvdQsQy6FZFCp4Utj1jAyIZsJuMuUFmkOyDxLYJQu1rcrzqE4sdf48c9RuM4_A9j58s_uSJaXmgxb8O-Hg1XEbtvQ1d (accessed on 9 April 2023).
55. Saudi Arabia Announces 24-Hour Curfew for Eid Al-Fitr Holiday. Arab News. Available online: https://www.arabnews.com/node/1673736/saudi-arabia (accessed on 8 April 2023).
56. Reuters Staff Saudi Arabia to Introduce 24-Hour Daily Curfew across the Kingdom for Eid Holiday. Reuters. Available online: https://www.reuters.com/article/health-coronavirus-saudi-curfew/saudi-arabia-to-introduce-24-hour-daily-curfew-across-the-kingdom-for-eid-holiday-idUSL8N2CU7AC (accessed on 8 April 2023).
57. Almeneessier, A.S.; Bahammam, A.S. How Does Diurnal Intermittent Fasting Impact Sleep, Daytime Sleepiness, and Markers of the Biological Clock? Current Insights. *Nat. Sci. Sleep* **2018**, *10*, 439–452. [CrossRef]
58. Osman, F.; Haldar, S.; Henry, C.J. Effects of Time-Restricted Feeding during Ramadan on Dietary Intake, Body Composition and Metabolic Outcomes. *Nutrients* **2020**, *12*, 2478. [CrossRef]
59. Akbari, H.A.; Yoosefi, M.; Pourabbas, M.; Weiss, K.; Knechtle, B.; Vancini, R.L.; Trakada, G.; Saad, H.B.; Lavie, C.J.; Ghram, A. Association of Ramadan Participation with Psychological Parameters: A Cross-Sectional Study during the COVID-19 Pandemic in Iran. *J. Clin. Med.* **2022**, *11*, 2346. [CrossRef]
60. Renzo, L.D.; Gualtieri, P.; Pivari, F.; Soldati, L.; Attinà, A.; Cinelli, G.; Esposito, E.; Leggeri, C.; Caparello, G.; Barrea, L.; et al. Eating Habits and Lifestyle Changes during COVID-19 Lockdown: An Italian Survey. *J. Transl. Med.* **2020**, *18*. [CrossRef] [PubMed]
61. Baquerizo-Sedano, L.; Chaquila, J.A.; Aguilar, L.; Ordovás, J.M.; González-Muniesa, P.; Garaulet, M. Anti-COVID-19 Measures Threaten Our Healthy Body Weight: Changes in Sleep and External Synchronizers of Circadian Clocks during Confinement. *Clin. Nutr.* **2021**, *41*, 2988–2995. [CrossRef] [PubMed]
62. Alshahrani, S.M.; Alghannam, A.F.; Taha, N.; Alqahtani, S.S.; Al-Mutairi, A.; Al-Saud, N.; Alghnam, S. The Impact of COVID-19 Pandemic on Weight and Body Mass Index in Saudi Arabia: A Longitudinal Study. *Front. Public Health* **2022**, *9*, 775022. [CrossRef] [PubMed]
63. Mason, T.B.; Barrington-Trimis, J.; Leventhal, A.M. Eating to Cope With the COVID-19 Pandemic and Body Weight Change in Young Adults. *J. Adolesc. Health* **2021**, *68*, 277–283. [CrossRef] [PubMed]
64. The World Bank Group Individuals Using the Internet (% of Population)-Saudi Arabia. Online Resource, by The World Bank. Available online: https://data.worldbank.org/indicator/IT.NET.USER.ZS?end=2019&locations=SA&start=200%200 (accessed on 12 September 2023).

nutrients

MDPI

Systematic Review

The Impact of the COVID-19 Pandemic on Binge Eating Disorder: A Systematic Review

Alice Caldiroli [1], Davide La Tegola [1], Francesca Manzo [2], Alberto Scalia [2], Letizia Maria Affaticati [2], Enrico Capuzzi [1], Fabrizia Colmegna [1], Marios Argyrides [3], Constantinos Giaginis [4], Leonardo Mendolicchio [5], Massimiliano Buoli [6,7], Massimo Clerici [1,2] and Antonios Dakanalis [1,2,*]

1 Department of Mental Health, Fondazione IRCSS San Gerardo dei Tintori, Via G.B. Pergolesi 33, 20900 Monza, Italy; alice.caldiroli@irccs-sangerardo.it (A.C.); davide.lategola@irccs-sangerardo.it (D.L.T.); e.capuzzi1@campus.unimib.it (E.C.); fabrizia.colmegna@irccs-sangerardo.it (F.C.); massimo.clerici@unimib.it (M.C.)
2 Department of Medicine and Surgery, University of Milano Bicocca, Via Cadore 38, 20900 Monza, Italy; f.manzo5@campus.unimib.it (F.M.); a.scalia5@campus.unimib.it (A.S.); letizia.affaticati@gmail.com (L.M.A.)
3 Department of Psychology, Neapolis University Pafos, Paphos 8042, Cyprus; m.argyrides.1@nup.ac.cy
4 Department of Food Science and Nutrition, School of Environment, University of Aegean, 81400 Myrina, Greece; cgiaginis@aegean.gr
5 Istituto Auxologico Italiano IRCCS, U.O. dei Disturbi del Comportamento Alimentare, Ospedale San Giuseppe, 28824 Piancavallo, Italy; l.mendolicchio@auxologico.it
6 Department of Pathophysiology and Transplantation, University of Milan, Via Festa del Perdono 7, 20122 Milan, Italy; massimiliano.buoli@unimi.it
7 Department of Neurosciences and Mental Health, Fondazione IRCCS Ca' Granda Ospedale Maggiore Policlinico, Via F. Sforza 35, 20122 Milan, Italy
* Correspondence: antonios.dakanalis@unimib.it; Tel.: +39-3755-651904

Abstract: The aim of this systematic review was to synthesise the impact of the COVID-19 pandemic on binge eating disorder (BED) the new onset and course. Inclusion criteria: original articles and BED diagnosis; and the main outcomes: relationships between the COVID-19 pandemic and the new onset/clinical changes in BED, and specific results for BED. Exclusion criteria: mixed/inaccurate diagnoses and articles not written in English. We searched four databases and one registry until 5 May 2023. The quality appraisal was conducted using the Effective Public Health Practice Project (EPHPP) tool. Twelve studies with 4326 participants were included. All studies were observational with nine cross-sectional and three longitudinal. Four of the included studies investigated new-onset BED, while eight examined the BED clinical course of patients with a previous diagnosis. With the exception of one study, the available literature indicates both an increase in BED diagnoses and a clinical worsening during COVID-19. Major limitations include study quality (weak-to-moderate) and high heterogeneity in terms of pandemic phase, population, geographical areas, and psychometric tools. Our findings indicate that BED patients are particularly vulnerable to events characterised by social distancing and over-worry, and should be, therefore, carefully monitored. Further studies are needed to corroborate our findings, implement preventive strategies, and promote personalised treatments. PROSPERO registration number: CRD42023434106

Keywords: binge eating disorder; COVID-19 pandemic; coronavirus; lockdown; eating disorders

1. Introduction

The worldwide spread of SARS-Cov-2 in December 2019 led the World Health Organization (WHO) to declare that the coronavirus disease (COVID-19) was a pandemic on 11 March 2020 [1]. After the introduction of restrictive measures to reduce the infection (e.g., lockdown), a number of subjects experienced a worsening of mental well-being as a result of fear of contagion and social distancing [2,3].

Several authors reported critical impacts of the COVID-19 pandemic on eating disorders (EDs). In particular, three systematic reviews [4–6] and one meta-analysis [7] revealed an increased prevalence or re-exacerbation of EDs with an increase in related hospitalisations during lockdowns. Nevertheless, differences between the different diagnostic entities were rarely explored, and contrasting results on anorexia nervosa (AN) and bulimia nervosa (BN) were described. Some studies reported more binge eating (BE) episodes after lockdowns in patients affected by BN [8], while other authors failed to identify a significant impact of the pandemic on the body mass index (BMI) of subjects suffering from AN and BN [6,7]. Curiously, the effects of the COVID-19 pandemic on individuals diagnosed with binge eating disorder (BED) have been poorly explored till now, as more attention has been given to trans-diagnostic BE attitudes.

A BED is defined by the presence of objective BE episodes concomitantly with a sense of lack of control and without inappropriate compensatory behaviours [9]. Although the 11th revision of the International Classification of Diseases (ICD-11) [10] included the subjective binges in BED diagnostic criteria, it has been demonstrated that there is an agreement between ICD-11 and DSM-5 for the same ED diagnosis in 98% of the cases [11]. Of note, BED represents the most common ED, with a lifetime prevalence of 3.5% among women and 2% among men [12,13]. In females under 30 years of age, the prevalence varies between 0.6% and 6.1%, according to the country of origin [14].

A specific BE behaviour, as well as BE episodes, is also characteristic of other clinical conditions such as BN, emotional eating, or night eating syndrome. Nevertheless, in the 5th edition of the Diagnostic and Statistical Manual of Mental Disorders (DSM-5), BED was recognised as a specific condition, finally acquiring its own nosographic dignity [9]. In fact, BED is characterised by specific clinical features such as emotion dysregulation, impulsivity, and cognitive distortion, and underpinned by peculiar neurobiological mechanisms [15]. Moreover, BED is often concomitant with medical (e.g., metabolic syndrome) and psychiatric comorbidities, which together worsen the BED course and impair the quality of life of the subjects suffering from this condition [16,17].

At the same time, the persistence of BED symptoms can be enhanced by social and environmental factors [18]. For example, family-related adverse life experiences (ALEs), especially early loss of caregivers, family separations, and detrimental parental interactions, negatively impact BN and BED patients [19]. Moreover, Forester and colleagues [20] demonstrated that specific times of the day (in particular, dinner time, but also lunch time and late evening) favour binge eating in BED patients, being the periods with the strongest food cravings and difficulties in emotional regulation.

Focusing on the pandemic period, some authors identified risk/protective factors for the new onset of BE symptoms (not BED diagnosis) [21], including both European and United States (US) data [22]. For example, having a mental disorder in comorbidity, the presence of pandemic-associated worry, self-reported pandemic-related loneliness, and being female and belonging to an ethnic minority were all identified as risk factors for the onset or exacerbation of BED [21,22]. In contrast, suffering from the COVID-19 illness, a change in living situation, being a key worker, or being in treatment emerged as protective factors against BE onset/worsening during the restrictions [21].

The COVID-19 pandemic was a dramatic event and, especially the confined, represented a challenging environment for individuals with BED. To date, it remains unclear whether BED psychopathology benefitted from lockdown measures or became worse, and evidence regarding the impacts of the COVID-19 pandemic on patients with the BED is still scarce. In light of these considerations, the objective of the present review was to systematically summarise the current literature about the impacts of the COVID-19 pandemic on the new onset and course of the BED.

2. Methods

2.1. Search Strategy

In order to find relevant articles, a search in the main databases (PubMed, Web of Science, PsychInfo), as well as consultation in US registries (National Institutes of Health, ClinicalTrials.gov), was performed. The search was conducted from 1987 to 5 May 2023, using the keywords "((COVID-19) OR (SARS-Cov-2) OR (coronavirus) OR (lockdown)) AND ((binge eating) OR (binge eating disorder))". No filters or limits were used.

2.2. Eligibility Criteria

Inclusion criteria were (1) original articles; (2) reported diagnosis of BED; (3) when the main topic of the article focused on the relationship between the COVID-19 pandemic and new onset or clinical changes in the BED; and (4) when different diagnoses were considered only papers reporting results on the BED subgroup were taken into account. Exclusion criteria were (1) reviews, meta-analyses, commentaries, letters, case reports, pooled analyses, comments, case studies, and study protocols; (2) when there were mixed or not accurately described diagnoses (e.g., BE behaviour); or (3) when articles were not written in the English language.

2.3. Study Selection

Two authors searched and screened the titles and abstracts of each record independently. Potentially eligible records were flagged, and the full text was obtained. Full texts were then reviewed for final inclusion. A third author supervised the process and resolved doubts and disagreements.

2.4. Data Extraction

Two authors extracted relevant data independently; doubts and disagreements were resolved by a third author. The following information was extracted: paper author and title, publication year, time period during which the study was conducted, geographic area where the research was conducted, study design, population, inclusion and exclusion criteria, sample size, mean age, data on BMI, psychometric tools, and main findings. If relevant data were not reported in the selected articles, the corresponding author was contacted to obtain further information.

2.5. Quality Appraisal

Two reviewers assessed the quality of each included study using the "Quality Assessment Tool for Quantitative Studies" developed by the Effective Public Health Practice Project (EPHPP) [23]. The tool evaluates the following domains: selection bias, study design, confounders, blinding, data collection method, withdrawals, and dropouts. Each domain is rated as "strong", "moderate" or "weak". Global quality is then calculated as strong (no weak ratings), moderate (one weak rating), or weak (two or more weak ratings). A third author supervised the process and resolved any discrepancies.

2.6. Data Synthesis

Relevant data were tabulated. The following information was presented: author/publication year, period of interest, geographic area where the research was conducted, study design, population, inclusion and exclusion criteria, sample size, mean age, data on BMI, psychometric tools, main findings, and quality score. We also provided a narrative synthesis of the studies' characteristics and main findings.

This systematic review, including the search strategy and inclusion/exclusion criteria, followed the Preferred Reporting Items for Systematic Reviews and Meta-Analyses (PRISMA) guidelines [24,25]. The protocol was pre-registered on PROSPERO (CRD42023434106).

3. Results

3.1. Study Selection

The search of all the databases provided a total of 6728 citations. Among these, 12 were identified as duplicates. The number of records marked as ineligible by the automation tools was 6582. Of the 134 records screened, 32 studies were discarded because, after reviewing the abstracts, the papers dealt with another topic. Reviews and meta-analyses (n = 18), as well as case reports, case series, surveys, letters and commentaries (n = 4), were excluded. The full texts of the remaining 80 citations were examined in more detail, concluding that 68 studies did not meet the inclusion criteria because 38 did not provide specific results on the BED, and 30 studies analysed subjects without an accurate diagnosis of the BED. Twelve studies met the inclusion criteria and were included in the present review (Figure 1).

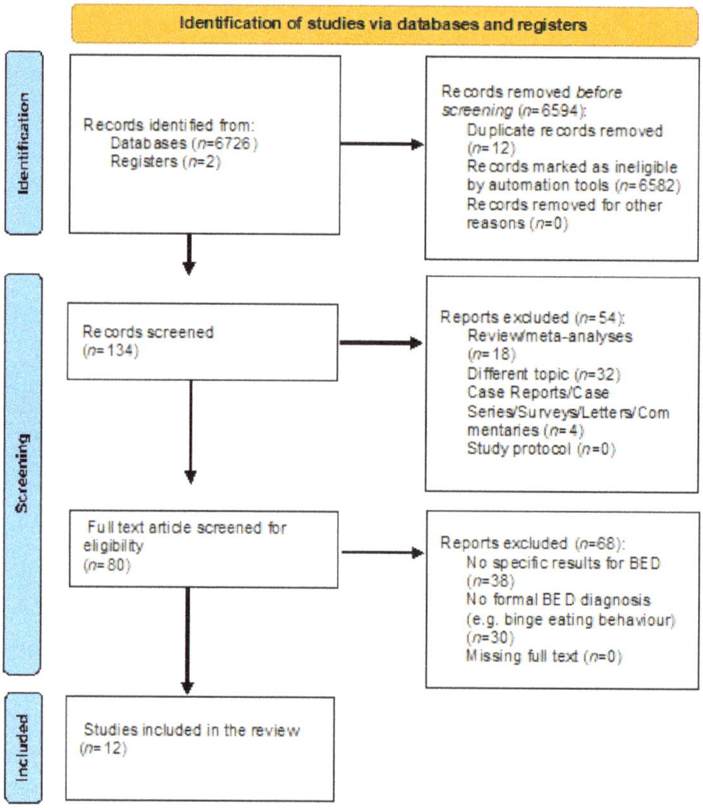

Figure 1. PRISMA diagram for systematic reviews.

3.2. Study Characteristics

The characteristics of each study were summarised in Tables 1–3. A total of 4326 participants from 12 studies were included in the present review. All studies were observational. Nine had a cross-sectional design [26–34], two had a longitudinal retrospective design [35,36], and one had a longitudinal prospective design [37]. The mean ages ranged from 19.3 to 45.5 years. Most studies were conducted in the US, Brazil, or in a European country. Four studies investigated the new onset of BED during the COVID-19 pandemic. Eight studies assessed pandemic-driven changes in the clinical course of patients with a previous diagnosis of the BED.

Table 1. Summary of findings about onset of Binge Eating Disorder during COVID-19 pandemic.

Study	Study Period	Country	Study Design	Data Collection Method/Psychometric Tools	Population	IC, EC	Sample Size (Females; Males; Non-Binary)	Mean Age (±SD)	BMI (at the Time of Study Initiation)	Main Findings	Quality Score [1]
Pimentel et al., 2021 [26]	June 2020 – October 2020	Brazil	Cross-sectional	Online questionnaire: EDS-7	HPs	IC: age 18–60; worker in a Brazilian hospital involved in the fight against COVID-19 EC: diagnosis of ED	219 (183; 36; 0)	35.6 (±13.6)	Underweight: 6 (2.7%) Normal weight: 104 (47.5%) Overweight: 71 (32.4%) Obesity grade I: 29 (13.2%) Obesity grade II: 8 (3.7%) Obesity grade III: 1 (0.5%)	- BED-related symptoms: 35 (16%) - BED: F (85.7%) > M - Association between BED and BMI ($p = 0.03$)	Moderate (weak data collection methods)
Garcês et al., 2022 [27]	23 October 2020 – 3 December 2020	Brazil	Cross-sectional, retrospective	Online questionnaires: BES and IPAQ Answers related to 2 times points: (1) pre-pandemic period; (2) during social distancing, i.e., 23 October 2020–03 December 2020	General population	IC: age 18–60; agreeing to the digital ICF). EC: refusal to agree to the ICF; children, adolescents, and elderly; incomplete answers and/or non-completion of questionnaires.	323 (218; 105; 0)	30.24	Normal weight: 194 (60.1%) Overweight or obese: 129 (39.9%)	- ↑ BE predicted body mass gain ($p < 0.001$) during social distancing - ↑ prevalence of BED both in the normal weight and in the overweight/obese group during social distancing ($p < 0.001$)	Moderate (weak data collection methods)
Bronfman and Chao, 2023 [28]	February 2021 – April 2021	US	Cross-sectional	Online questionnaire: EDE-Q, EPII, GAD-7, PHQ-8, SMUI and questions based on the YRBSQ	College students	IC: college students in the US; age 18–23; completion of depression and EDs questionnaires. EC: N/A.	202 (168; 30; 4)	20.7 (±1.7)	Mean BMI: 23.2 (±4.2)	- BED: 27.2% - ↑ depressive symptoms = ↑ odds of BED ($p < 0.001$) - ↑ anxiety symptoms = ↑ odds of BED ($p < 0.001$) - ↑ social media use = ↑ odds of BED ($p = 0.03$)	Moderate (weak selection bias)

Table 1. *Cont.*

Study	Study Period	Country	Study Design	Data Collection Method/Psychometric Tools	Population	IC, EC	Sample Size (Females; Males; Non-Binary)	Mean Age (±SD)	BMI (at the Time of Study Initiation)	Main Findings	Quality Score [1]
Murray et al., 2023 [37]	1 January 2017 – 31 December 2020	US	Longitudinal prospective	DMSS; PHA: MHA questions	Military Members (in service)	IC: active component service members from all military services from 1 January 2017 to 31 December 2021. Answer "yes" to questions on "major life stressors", "history of mental health care", "PTSD screening" and "depression screening" EC: hospitalisation or visits for an ED diagnosis prior to the surveillance period	2454 ED diagnoses (1556; 898; 0)	N/A	N/A	-BED annual incidence rate = 0.7 per 10.000 p-yrs -From the first year of the surveillance period to the endpoint, BED annual incidence rates tripled	Moderate (weak selection bias)

Legend: [1] according to [23]; ↑ = increased; BE = binge-eating; BED = binge eating disorder; BEDS-7 = 7-item binge eating disorder screener; BES = binge eating scale; BMI = body mass index; DMSS = defence medical surveillance system; EC = exclusion criteria; EDs = eating disorders; EDE-Q = Eating Disorders Examination Questionnaire; EPII = epidemic-pandemic impact inventory; F = females; GAD-7 = generalised anxiety disorder-7 scale; HPs = health professionals; IC = inclusion criteria; ICF = informed consent form; IPAQ = short version of the International Physical Activity Questionnaire; M = males; MHA = mental health assessment; p = p value; N/A = not available; PHA = periodic health assessment; PHQ-8 = patient health questionnaire; p-yrs = per years; PTSD = post-traumatic stress disorder; SD = standard deviation; SMUI = social media usage index; US = United States; YRBSQ = Youth Risk Behaviour Survey Questionnaire.

Table 2. Summary of findings from moderate-quality studies on BED clinical changes/relapse during COVID-19 pandemic.

Study	Study Period	Country	Study Design	Data Collection Method/Psychometric Tools	Population	IC, EC	Sample Size (Females; Males; Other)	Mean Age (±SD)	BMI (at the Time of Study Initiation)	Main Findings	Quality Score [1]
Termorshuizen et al., 2020 [29]	April 2020 – May 2020	NL and US	Cross-sectional	Online survey: quantitative measures and free-text responses investigating the impact of COVID-19 on eating disorders, general physical and mental well-being (GAD-7), treatment of EDs	Individuals with self-reported ED	IC: minimum age 18 in US, minimum age 26 in NL; life-time self-reported ED EC: N/A	1021 (1001; 18; 2) 216 affected by BED	US: 30.61 (±9.37) NL: N/A	N/A	- ↑ BE episodes and urges to binge in BN and BED - BE of stockpiled food: BED and BN > AN ($p < 0.001$)	Moderate (weak data collection methods)
Giel et al., 2021 [35]	Spring 2020	Germany	Longitudinal retrospective Comparison between 4 weeks immediately before lockdown (on a recall basis) and 4 weeks during lockdown	BE episodes: EDE Other assessments: BDI-II, ERQ, PSQ, SCID SOC-L9	BED patients (who had previously participated in the IMPULS trial, investigating a novel psychotherapy for BED)	IC: BED diagnosis according to DSM-5; minimum age 18 years. EC: suicidality, pregnancy or lactation, a severe mental or somatic disorder	42 (34; 8; 0)	45.5 (±12.6)	Mean BMI: 34.0 (±6.9)	↑ BE episodes in the 4 weeks during lockdown vs 4 weeks before lockdown ($p < 0.001$)	Moderate (weak confounder)

Table 2. *Cont.*

Study	Study Period	Country	Study Design	Data Collection Method/Psychometric Tools	Population	IC, EC	Sample Size (Females; Males; Other)	Mean Age (±SD)	BMI (at the Time of Study Initiation)	Main Findings	Quality Score [1]
Baenas et al., 2022 [36]	August 2020 – January 2021	Europe (Austria, Czech Republic, Germany, Lithuania, Portugal, Russia, Spain, Ukraine) and Asia (China, Korea, Japan)	Longitudinal retrospective Comparison pre- vs. post-lockdown	CIES	EDs patients	IC: ED diagnosis according to DSM-5 criteria, performed by expert clinical psychologists and psychiatrists using a semi-structured clinical interview (SCID-5). EC: N/A	829 (584; 245; 0) 113 affected by BED	27.9 (±12.3).	BED Mean BMI: 35.08 (±9.63)	- Patients affected by BED experienced the highest impact on weight and clinical symptoms in comparison with other EDs during lockdown. -BED patients during post- vs pre-lockdown: --↑ weight ($p < 0.001$; Cohen's d = 0.12) -- ↑ BMI ($p < 0.001$; Cohen's d = 0.14) -- impaired eating style ($p = 0.01$; Cohen's d = 0.18) -- ↑ anxiety-depression symptoms ($p = 0.004$; Cohen's d = 0.16)	Moderate (weak confounder)
Baenas et al., 2023 [34]	June 2020 – March 2021	Brazil, Portugal, Spain	Cross-sectional (retrospectively collected data) Comparison post- vs pre-lockdown	CIES (subscales I, II, III)	EDs patients	IC: females; ED diagnosis according to DSM-5 EC: N/A	264 (264; 0; 0) 81 affected by BED	33.49 (±12.54)	BED Mean BMI: 33.70 (±7.37)	BED patients during post- vs pre-lockdown: - ↑ weight ($p < 0.001$; Cohen's d = 0.18) - ↑ BMI ($p < 0.001$; Cohen's d = 0.20) - ↑ anxiety-depression ($p < 0.001$; Cohen's d = 0.73) - ↑ emotional dysregulation ($p < 0.001$; Cohen's d = 0.28)	Moderate (weak selection bias)

Legend: [1] according to [23]; ↑ = increased; AN = anorexia nervosa; BDI-II = Beck Depression Inventory-II; BE = binge-eating; BED = binge eating disorder; BMI = body mass index; BN = bulimia nervosa; CIES = COVID Isolation Eating Scale; DSM = Diagnostic and Statistical Manual of Mental Disorders; EC = exclusion criteria; EDE = eating disorder examination; EDs = eating disorders; ERQ = Emotion Regulation Questionnaire; F = females; GAD-7 = Generalised Anxiety Disorder-7 scale; IC = inclusion criteria; M = males; NL = Netherlands; p = p value; PSQ = Perceived Stress Questionnaire; SCID = Structured Clinical Interview for DSM-5; SOC-L9 = Sense of Coherence Scale-Leipzig 9 items; US = United States; vs = versus.

Table 3. Summary of findings from weak-quality studies on BED clinical changes/relapse during COVID-19 pandemic.

Study	Study Period	Country	Study Design	Data Collection Method/Psychometric Tools	Population	IC, EC	Sample Size (Females; Males; Other)	Mean Age (±SD)	BMI (at the Time of Study Initiation)	Main Findings	Quality Score [1]
Colleluori et al., 2021 [30]	9 March 2020 – 8 May 2020 (first lockdown)	Italy	Cross-sectional	Ad hoc online survey investigating the experience of HCPs in the management of EDs (AN, BN, BED) during phase I lockdown	Patients with DSM-5-defined EDs (AN, BN, BED)	IC: DSM-5-defined AN, BN, BED EC: other ED (e.g., night eating syndrome, vigorexia or ortorexia)	453 (N/A) 174 affected by BED	N/A	N/A	- BE frequency: increased in 30.5% and reduced in 24.7% of patients affected by BED - ↑ BE episodes BN > BED ($p < 0.05$) - switching ED category: BN > BED or AN ($p = 0.001$) - ↑ visits frequency: AN = BN = BED ($p = 0.113$)	Weak (weak selection bias and weak data collection method)
De Pasquale et al., 2021 [31]	March 2020 – February 2021	Italy	Cross-sectional	Online questionnaires: BES, EDI-2, FCV-19s, POMS	College students	N/A	469 (248; 221; 0)	22.47 (±2.70)	N/A	- Fear of COVID-19 → ↑ BES scores ($p < 0.01$) - BED: F > M ($p = 0.03$)	Weak (weak selection bias and weak data collection methods)
Frayn et al., 2021 [32]	August 2020	US	Cross-sectional	Individual, semi-structured interview (conducted by a trained researcher) investigating: - perceived impact of COVID-19 on ED symptoms prior to starting treatment - experience with teletherapy - feedback on adapting treatment to address COVID-related concerns	Patients with BE spectrum disorders (undergoing a therapy treatment program for BE spectrum disorders, i.e., COMPASS)	IC: age 18–70; at least 12 objective or subjective binge episodes in the past 3 months; meeting DSM-5 criteria for an ED EC: BMI < 16; already receiving ED treatment; requiring immediate treatment for medical complications resulting from ED symptoms; severe psychopathology that would limit their ability to comply (e.g., severe depression with suicidal intent, active psychotic disorder)	11 (7; 3; 1) 7 affected by BED	42.8 (±14.2)	Mean BMI: 34.7 (±10.3)	Qualitative findings: -both symptom worsening and improvement during COVID-19 -Social distancing and stay-at-home measures were found to both improve and worsen symptoms for different patients	Weak (weak selection bias and weak data collection methods)

Table 3. *Cont.*

Study	Study Period	Country	Study Design	Data Collection Method/Psychometric Tools	Population	IC, EC	Sample Size (Females; Males; Other)	Mean Age (±SD)	BMI (at the Time of Study Initiation)	Main Findings	Quality Score [1]
Freizinger et al., 2022 [33]	November 2020 – January 2021	Northeastern US	Cross-sectional (retrospectively collected data: questions pertained to the initial lockdown period of the COVID-19 pandemic)	Online survey: BES, CEFIS-AYA	Patients referred for evaluation and treatment for BED, emotional eating, or lifestyle counselling for weight concerns and/or preparation for bariatric surgery	IC: age 13–26; evaluated and/or treated from March 2019 to March 2020. EC: N/A	39 (31; 6; 2) 22 affected by BED	19.3 (±3.2)	N/A	- ↑ COVID-related stress = ↑ BES scores ($p = 0.04$) - ↓ age = ↑ BES scores ($p = 0.02$) - No association between BES scores and: -- food availability at home ($p = 0.97$) -- worriedness about having sufficient food in the home ($p = 0.40$) -- difficulty obtaining food ($p = 0.84$)	Weak (weak selection bias and data collecting method)

Legend: [1] according to [23]; ↑ = increased; ↓ = decreased; AN = anorexia nervosa; BE = binge-eating; BED = binge eating disorder; BES = binge eating scale; BMI = body mass index; BN = bulimia nervosa; CEFIS-AYA= COVID-19 Exposure and Family Impact Survey-Adolescent and Young Adult Version; DSM = Diagnostic and Statistical Manual of Mental Disorders; EC = exclusion criteria; EDI-2 = eating disorder inventory-2; EDs = eating disorders; F = females; FCV-19S = fear of COVID-19 scale; HCPs = healthcare providers; IC = inclusion criteria; M = males; p = p value; POMS = profile of mood states; US = United States.

3.3. Quality Appraisal

The quality appraisal was conducted using the "Quality Assessment Tool for Quantitative Studies" developed by the EPHPP [23]. Two-thirds of the included studies were rated as "moderate" in quality [26–29,34–37] (Tables 1 and 2); four studies were rated as "weak" [30–33] (Table 3). None of the studies received a "strong" quality rating.

3.4. BED New Onset during COVID-19 Pandemic

We included four studies about the new onset of the BED during the COVID-19 pandemic. Among them, two were performed on the Brazilian population [26,27] and two on the US population [28,37].

Pimentel and his group [26] published a prospective, cross-sectional study assessing the relationship between the COVID-19 pandemic and the occurrence of BED and mental suffering in a sample of Brazilian health professionals. The online interview was completed by 219 professionals, of which 35 (16%) reported BED-related symptoms. This study also found a higher prevalence of BED in women than men and a direct association between BED onset and high BMI. Similarly, a cross-sectional and retrospective study was carried out by Garcês and colleagues [27], who compared 94 (60.1%) normal-weight adults and 129 (39.9%) overweight/obese subjects. The results of this study confirmed the increased risk of developing BED during social distancing both in the overweight/obese group and in normal-weight adults.

The negative impact of the COVID-19 pandemic on eating behaviour was also demonstrated by Bronfman and Chao [28] who examined, differently from the first two studies, a population of US college students between February and April 2021. In this cross-sectional study, factors potentially associated with ED symptoms were analysed, finding that individuals with more severe depressive symptoms had higher odds of meeting criteria for BN, BED, and subthreshold EDs, while more severe anxiety symptoms and more frequent social media use were associated with increased odds of BED.

Finally, an interesting longitudinal study by Murray and collaborators [37] was conducted to update annual incidence rates and demographic trends of EDs among military service members between 2017 and 2021. Their results demonstrated that the annual incidence rates of all EDs increased from 2017 to 2019, decreased in 2020, and increased again in 2021. From the first year of the surveillance period (2017) to the endpoint (2020), the BED annual incidence rate tripled.

All included studies investigating the relationship between the COVID-19 pandemic and BED new onset are detailed in Table 1.

3.5. BED Clinical Changes/Relapse during COVID-19 Pandemic

Eight studies were included about the impacts of the COVID-19 pandemic on BED clinical courses. Most authors considered the COVID-19 pandemic and related lockdowns as a traumatic event, which promoted dysfunctional eating behaviours.

Four studies that analysed the first lockdown period in different countries reported similar results. In the study by Termorshuizen et al. [29], 1021 subjects (216 with a self-reported diagnosis of BED) were recruited from the US and the Netherlands (NL) via social media. Lifetime diagnosis of a BED represented 35% of the US sample and 23% of the Dutch one. Both patients affected by BED and BN reported more frequent BE episodes, especially of stockpiled foods. In the second study, Giel et al. [35] investigated ED symptoms, as well as general psychopathology, in 42 patients with a history of a BED who previously sought treatment in the randomised–controlled IMPULS trial in Germany. The authors assessed self-reported ED pathology using the Eating Disorder Examination (EDE) and demonstrated an increase in BE episodes in the four weeks during lockdown compared to the four weeks before lockdown. The cross-sectional study by Freizinger et al. [33] analysed a sample of 22 adolescents and young adults affected by BED. The authors used the binge eating scale (BES), which is a self-administered questionnaire assessing eating behaviours, feelings, and cognitive aspects linked to BE, and the COVID-19 Exposure and Family Impact

Survey-Adolescent and Young Adult version (CEFIS-AYA), which investigates experiences related to the COVID-19 pandemic. They found that the degree of COVID-related stress was associated with more BE behaviours. More recently, Baenas and colleagues [34] evaluated clinical changes in patients with EDs during the first lockdown in a sample of 81 female patients. Clinical information was collected using the COVID Isolation Eating Scale (CIES), a self-report questionnaire composed of four subscales investigating the circumstances of the quarantine, the effects of the lockdown on eating symptoms, the effects of the confinement on eating behaviours, and the impacts of remote interventions. The authors found that post-lockdown BED patients experienced weight gain and a worsening of anxiety, depression and emotional dysregulation. These findings confirmed previous research by the same group that had found similar results in a multicentre study using the same instrument (CIES) but focused on the second wave of the COVID-19 pandemic. A larger sample (N = 829) was recruited, including 113 individuals with a BED diagnosis, who reported the highest impact on weight and eating style impairment in comparison with other subgroups and an increase in anxiety–depressive symptoms [36].

De Pasquale and collaborators [31] investigated, in a sample of 469 Italian college students, the relationship between mood states and eating behaviours during the first year of the COVID-19 pandemic. The utilised online protocol included the following psychometric tools: the Fear of COVID-19 Scale (FCV-19S) for the evaluation of fear, the Profile of Mood States (POMS) for the assessment of emotional states, the Eating Disorder Inventory-2 (EDI-2) and the BES. A significant effect of fear of COVID-19 in worsening BE behaviours was identified by the authors.

On the other hand, Colleluori et al. [30], although analysing a large sample of BED patients (N = 174) during the same period (first lockdown), reported different findings. The data were retrieved from an online survey, investigating the experience of healthcare providers (HCPs) involved in the treatment of EDs. The objective of this study was to assess whether social restrictions represented a stressor for ED patients and their HCPs, if they affected therapeutic alliance and altered the frequency of dysfunctional behaviours. The authors found a high variability in the changes in frequency of BE episodes after the pandemic, which increased in 30.5% and decreased in 24.7% of BED subjects. Similarly, Frayn et al. [32] observed both symptom deterioration and improvement in the BED, analysing seven BED patients. Participants underwent a semi-structured interview and authors identified factors that may decrease BE, such as having greater control over the food environment.

A summary of the studies on the BED clinical course during COVID-19 is reported in Tables 2 and 3.

4. Discussion

Most studies evaluating the impacts of the COVID-19 pandemic on EDs, focused on unspecific BE behaviours or BE episodes that are symptoms shared by the different EDs and characterise other clinical conditions, such as obesity or emotional eating. BE episodes are core symptoms of BED [9,10], which represents a defined nosographic entity underpinned by specific neurobiological mechanisms [38,39]. To the best of our knowledge, this is the first systematic review that synthesises the impact of the COVID-19 pandemic and the related sanitary actions on the BED, diagnosed according to DSM-5 [9]. Focusing on the specific disorder, rather than unspecific binge eating symptoms, may shed light on some pathophysiological mechanisms underlying the BED and its etiopathogenesis, and may foster the implementation of targeted preventive/treatment strategies for subjects suffering from this specific ED.

Two main findings emerged from our systematic literature review:

1. the prevalence of the BED during the COVID-19 pandemic increased based on data collected from Brazil and the US;
2. the patients affected by the BED experienced a worsening of symptoms during the first year of the COVID-19 pandemic.

The four studies investigating the impacts of pandemic lockdowns on new diagnoses of BED found an increase in the prevalence of BED and its symptoms [26–28,37]. Interestingly, a direct association between the BED and BMI levels during the first pandemic wave was identified [26,27], as well as a higher prevalence of the BED in female than male HPs [26], in line with previous reports [40].

Six papers demonstrated a clinical worsening of the BED during the first lockdown as shown by more BE episodes [29,35], the elevation of the BMI, and more severe anxiety and depressive symptoms, as well as emotion dysregulation and impaired eating styles [34,36]. These findings confirm the available data in the literature [6,7,41,42], however, focused on unspecific BE behaviours regardless of BED diagnosis. In contrast, the paper by Machado et al. [43] found no significant change in symptoms and BMI after the lockdown confinement in a sample of patients suffering from EDs; however, only two patients had a diagnosis of BED, thus supporting the hypothesis that this group of subjects might suffer more from the negative consequences of social distancing. In addition, the two weak-quality studies outlined mixed results about the impacts of the pandemic on the clinical course of the BED [30,32]. It must be specified that the Italian study by Colleluori and colleagues [30] collected indirect information with online surveys addressed to HCPs.

The studies described in this review suggest that the higher frequency of the BED or the worsening of its course may be mediated by other factors influenced by the pandemic, such as the presence of depression/anxiety [28,34,36], fear of being infected [31], or an increase in BMI [26,34,36]. Of note, after the outbreak of the pandemic, BMI increased in the general population, especially among women and the youngest [44] which represent groups vulnerable to developing BED [45]. Similarly, symptoms such as depression and anxiety have also been frequently reported in the general population since the outbreak of the pandemic [46]. On the other hand, the presence of affective symptoms in patients affected by EDs may have potentiated the negative effects of social isolation and distancing on the clinical course of these individuals [47]. These observations suggest a role of neuropeptides, such as oxytocin, a molecule implicated in social behaviour [48] and appetite regulation [49,50], in explaining the complex entanglement of depressive symptoms, worsening of BED symptoms, weight gain, and social distancing. In support of this hypothesis, previous research suggested that the administration of intranasal oxytocin is able to reduce food intake and cravings [51].

Limitations

This systematic review presented different limitations. First of all, the relatively small sample size of the included studies, as well as the cross-sectional design of most papers, with self-reported clinical information, may have contributed to making the quality of studies predominantly moderate or weak. Second, the high heterogeneity of the included papers in terms of pandemic phase, population, geographical areas, and psychometric tools may reduce the power of the reported findings. The various phases of the pandemic were characterised by important differences in terms of the magnitude of restrictions, social distancing, and the use of telemedicine [52]. Moreover, the studies included in the present review were almost all conducted in Europe and America. Only one study [36] included a group of patients from Asia. Of note, the epidemiology of EDs differs across countries, being that eating behaviours are strongly influenced by the environment [53,54] and cultural factors, such as body image models proposed by the media [55]. In particular, the American continent has the highest prevalence for all EDs, followed by Asia and Europe, while data coming from Africa are scarce [56]. Finally, it should be noted that our results referred to very heterogeneous samples, including HPs, college students, and military service members, as well as the general population, limiting the generalizability of the results presented in this review.

5. Conclusions

In conclusion, our systematic review confirmed that social restrictions related to the COVID-19 pandemic have negatively impacted EDs, especially the BED. The strong entanglement between the restrictions, an elevated BMI, and the presence of depressive symptoms could have promoted the development or the worsening of the BED. The included studies presented some methodological weaknesses, making it difficult to draw definitive conclusions. Further research is necessary to clarify the underlying clinical and biological factors triggering the worsening of the BED when dramatic events occur, such as a pandemic. The final purpose is to foster more personalised interventions in order to prevent and promptly treat BED symptoms in predisposed and vulnerable subjects.

Author Contributions: Conceptualisation, A.C., D.L.T. and M.B.; methodology, A.C. and M.B.; validation, M.A., C.G., F.C. and L.M.; investigation, F.M. and A.S.; data curation, E.C.; writing—original draft preparation, A.C., D.L.T., F.M., A.S. and L.M.A.; writing—review and editing, A.C., L.M.A, E.C. and M.B.; visualisation, A.D.; supervision, M.B., M.C. and A.D.; project administration, F.C and L.M. All authors have read and agreed to the published version of the manuscript.

Funding: This research received no external funding.

Institutional Review Board Statement: Not applicable.

Informed Consent Statement: Not applicable.

Data Availability Statement: The datasets generated during and/or analysed during the current study are available from the corresponding author on reasonable request.

Conflicts of Interest: The authors declare no conflict of interest.

References

1. WHO Director-General's Opening Remarks at the Media Briefing on COVID-19—11 March 2020. Available online: https://www.who.int/director-general/speeches/detail/who-director-general-s-opening-remarks-at-the-media-briefing-on-covid-19{-}{-}-11-march-2020 (accessed on 9 July 2023).
2. Daly, M.; Robinson, E. Anxiety reported by US adults in 2019 and during the 2020 COVID-19 pandemic: Population-based evidence from two nationally representative samples. *J. Affect. Disord.* **2021**, *286*, 296–300. [CrossRef] [PubMed]
3. Caldiroli, A.; Capuzzi, E.; Tringali, A.; Tagliabue, I.; Turco, M.; Fortunato, A.; Sibilla, M.; Montana, C.; Maggioni, L.; Pellicioli, C.; et al. The psychopathological impact of the SARS-CoV-2 epidemic on subjects suffering from different mental disorders: An observational retrospective study. *Psychiatry Res.* **2022**, *307*, 114334. [CrossRef] [PubMed]
4. Miniati, M.; Marzetti, F.; Palagini, L.; Marazziti, D.; Orrù, G.; Conversano, C.; Gemignani, A. Eating Disorders Spectrum During the COVID Pandemic: A Systematic Review. *Front. Psychol.* **2021**, *12*, 663376. [CrossRef] [PubMed]
5. Gao, Y.; Bagheri, N.; Furuya-Kanamori, L. Has the COVID-19 pandemic lockdown worsened eating disorders symptoms among patients with eating disorders? A systematic review. *J. Public Health* **2022**, *30*, 2743–2752. [CrossRef] [PubMed]
6. Devoe, J.D.; Han, A.; Anderson, A.; Katzman, D.K.; Patten, S.B.; Soumbasis, A.; Flanagan, J.; Paslakis, G.; Vyver, E.; Marcoux, G.; et al. The impact of the COVID-19 pandemic on eating disorders: A systematic review. *Int. J. Eat. Disord.* **2022**, *56*, 5–25. [CrossRef] [PubMed]
7. Güzel, Â.; Mutlu, N.L.; Molendijk, M. COVID-19-related changes in eating disorder pathology, emotional and binge eating and need for care: A systematic review with frequentist and Bayesian meta-analyses. *Eat. Weight Disord.* **2023**, *28*, 19. [CrossRef] [PubMed]
8. Castellini, G.; Cassioli, E.; Rossi, E.; Innocenti, M.; Gironi, V.; Sanfilippo, G.; Felciai, F.; Monteleone, A.M.; Ricca, V. The impact of COVID-19 epidemic on eating disorders: A longitudinal observation of pre versus post psychopathological features in a sample of patients with eating disorders and a group of healthy controls. *Int. J. Eat. Disord.* **2020**, *53*, 1855–1862. [CrossRef] [PubMed]
9. American Psychiatric Association. *Diagnostic and Statistical Manual of Mental Disorders*, 5th ed.; American Psychiatric Press: Washington, DC, USA, 2013.
10. World Health Organization. International Classification of Diseases. 2022. Available online: https://icd.who.int/browse11/l-m/en#/http://id.who.int/icd/entity/1673294767 (accessed on 14 August 2023).
11. Quadflieg, N.; Voderholzer, U.; Meule, A.; Fichter, M.M. Comparing ICD-11 and DSM-5 eating disorder diagnoses with the Munich eating and feeding disorder questionnaire (ED-Quest). *Int. J. Eat. Disord.* **2023**, *ahead of print*. [CrossRef]
12. Kessler, R.C.; Berglund, P.A.; Chiu, W.T.; Deitz, A.C.; Hudson, J.I.; Shahly, V.; Aguilar-Gaxiola, S.; Alonso, J.; Angermeyer, M.C.; Benjet, C.; et al. The prevalence and correlates of binge eating disorder in the World Health Organization World Mental Health Surveys. *Biol. Psychiatry* **2013**, *73*, 904–914. [CrossRef]

13. Silén, Y.; Keski-Rahkonen, A. Worldwide prevalence of DSM-5 eating disorders among young people. *Curr. Opin. Psychiatry* **2022**, *35*, 362–371. [CrossRef]
14. Hay, P.; Aouad, P.; Le, A.; Marks, P.; Maloney, D.; National Eating Disorder Research Consortium; Touyz, S.; Maguire, S. Epidemiology of eating disorders: Population, prevalence, disease burden and quality of life informing public policy in Australia-a rapid review. *J. Eat. Disord.* **2023**, *11*, 23. [CrossRef] [PubMed]
15. Colton, E.; Wilson, K.E.; Chong, T.T.; Verdejo-Garcia, A. Dysfunctional decision-making in binge-eating disorder: A meta-analysis and systematic review. *Neurosci. Biobehav. Rev.* **2023**, *152*, 105250. [CrossRef] [PubMed]
16. Dakanalis, A.; Clerici, M. The nature of the association between binge-eating severity and metabolic syndrome. *Eat. Weight Disord.* **2017**, *22*, 553–554. [CrossRef] [PubMed]
17. Udo, T.; Grilo, C.M. Prevalence and correlates of DSM-5-defined eating disorders in a nationally representative sample of U.S. adults. *Biol. Psychiatry* **2018**, *84*, 345–354. [CrossRef] [PubMed]
18. Hübel, C.; Marzi, S.J.; Breen, G.; Bulik, C.M. Epigenetics in eating disorders: A systematic review. *Mol. Psychiatry* **2019**, *24*, 901–915. [CrossRef] [PubMed]
19. Grogan, K.; MacGarry, D.; Bramham, J.; Scriven, M.; Maher, C.; Fitzgerald, A. Family-related non-abuse adverse life experiences occurring for adults diagnosed with eating disorders: A systematic review. *J. Eat. Disord.* **2020**, *8*, 36. [CrossRef] [PubMed]
20. Forester, G.; Schaefer, L.M.; Dodd, D.R.; Burr, E.K.; Bartholomay, J.; Berner, L.A.; Crosby, R.D.; Peterson, C.B.; Crow, S.J.; Engel, S.G.; et al. Time-of-day and day-of-week patterns of binge eating and relevant psychological vulnerabilities in binge-eating disorder. *Int. J. Eat. Disord.* **2023**. *ahead of print*. [CrossRef] [PubMed]
21. Davies, H.L.; Hübel, C.; Herle, M.; Kakar, S.; Mundy, J.; Peel, A.J.; ter Kuile, A.R.; Zvrskovec, J.; Monssen, D.; Lim, K.X.; et al. Risk and protective factors for new-onset binge eating, low weight, and self-harm symptoms in >35,000 individuals in the UK during the COVID-19 pandemic. *Int. J. Eat. Disord.* **2023**, *56*, 91–107. [CrossRef]
22. Thompson, K.A.; Hedlund, E.L.; Sun, Q.; Peat, C.M.; Goode, R.W.; Termorshuizen, J.D.; Thornton, L.M.; Borg, S.; van Furth, E.F.; Birgegård, A.; et al. Course and predictors of eating disorder symptoms, anxiety symptoms, and pandemic-related eating disorder concerns among adults with eating disorders during the first year of the COVID-19 pandemic. *Int. J. Eat. Disord.* **2023**, *56*, 151–168. [CrossRef]
23. Thomas, B.H.; Ciliska, D.; Dobbins, M.; Micucci, S. A process for systematically reviewing the literature: Providing the research evidence for public health nursing interventions. *Worldviews Evid. Based Nurs.* **2004**, *1*, 176–184. [CrossRef]
24. Moher, D.; Liberati, A.; Tetzlaff, J.; Altman, D.G.; PRISMA Group. Preferred reporting items for systematic reviews and meta-analyses: The PRISMA statement. *PLoS Med.* **2009**, *6*, e1000097. [CrossRef] [PubMed]
25. Page, M.J.; McKenzie, J.E.; Bossuyt, P.M.; Boutron, I.; Hoffmann, T.C.; Mulrow, C.D.; Shamseer, L.; Tetzlaff, J.M.; Akl, E.A.; Brennan, S.E.; et al. The PRISMA 2020 statement: An updated guideline for reporting systematic reviews. *BMJ* **2021**, *372*, n71. [CrossRef] [PubMed]
26. Pimentel, R.F.W.; Rodrigues, L.M.; Rocha, L.R.; Santana, A.I.C.; de Figueiredo, P.C.M.; Carvalho, M.L.V.; da Silva, D.A.R.; Suen, V.M.M.; Merces, M.C.D. Relationship between the COVID-19 pandemic, binge eating, and mental suffering in health professionals in Brazil: A cross-sectional study. *Rev. Bras. Med. Trab.* **2021**, *19*, 283–289. [CrossRef]
27. Garcês, C.P.; Oliveira, E.; Silva, L.; Nunes, S.M.; Cheik, N.C. Effects of social distancing caused by the COVID-19 pandemic on physical activity level, sitting time, and binge eating: A comparison between overweight/obese and normal-weight adults. *Sport Sci. Health* **2022**, *18*, 1505–1512. [CrossRef] [PubMed]
28. Bronfman, K.; Chao, A.M. Psychosocial Factors, Technology Use, and Disordered Eating in College Students During the COVID-19 Pandemic. *J. Psychosoc. Nurs. Ment. Health Serv.* **2023**, *ahead of print*. 1–10. [CrossRef] [PubMed]
29. Termorshuizen, J.D.; Watson, H.J.; Thornton, L.M.; Borg, S.; Flatt, R.E.; MacDermod, C.M.; Harper, L.E.; Furth, E.F.; Peat, C.M.; Bulik, C.M. Early impact of COVID-19 on individuals with self-reported eating disorders: A survey of ~1000 individuals in the United States and the Netherlands. *Int. J. Eat. Disord.* **2020**, *53*, 1780–1790. [CrossRef] [PubMed]
30. Colleluori, G.; Goria, I.; Zillanti, C.; Marucci, S.; Ragione, L.D. Eating disorders during COVID-19 pandemic: The experience of Italian healthcare providers. *Eat. Weight Disord.* **2021**, *26*, 2787–2793. [CrossRef] [PubMed]
31. De Pasquale, C.; Sciacca, F.; Conti, D.; Pistorio, M.L.; Hichy, Z.; Cardullo, R.L.; Di Nuovo, S. Relations Between Mood States and Eating Behavior During COVID-19 Pandemic in a Sample of Italian College Students. *Front. Psychol.* **2021**, *12*, 684195. [CrossRef]
32. Frayn, M.; Fojtu, C.; Juarascio, A. COVID-19 and binge eating: Patient perceptions of eating disorder symptoms, tele-therapy, and treatment implications. *Curr. Psychol.* **2021**, *40*, 6249–6258. [CrossRef]
33. Freizinger, M.; Jhe, G.B.; Dahlberg, S.E.; Pluhar, E.; Raffoul, A.; Slater, W.; Shrier, L.A. Binge-eating behaviors in adolescents and young adults during the COVID-19 pandemic. *J. Eat. Disord.* **2022**, *10*, 125. [CrossRef]
34. Baenas, I.; Neufeld, C.B.; Ramos, R.; Munguía, L.; Pessa, R.P.; Rodrigues, T.; Jiménez-Murcia, S.; Gonçalves, S.; Teodoro, M.C.; Pinto-Bastos, A.; et al. Eating disorders during lockdown: The transcultural influence on eating and mood disturbances in Ibero-Brazilian population. *J. Eat. Disord.* **2023**, *11*, 39. [CrossRef] [PubMed]
35. Giel, K.E.; Schurr, M.; Zipfel, S.; Junne, F.; Schag, K. Eating behaviour and symptom trajectories in patients with a history of binge eating disorder during COVID-19 pandemic. *Eur. Eat. Disord. Rev.* **2021**, *29*, 657–662. [CrossRef] [PubMed]
36. Baenas, I.; Etxandi, M.; Munguía, L.; Granero, R.; Mestre-Bach, G.; Sánchez, I.; Ortega, E.; Andreu, A.; Moize, V.L.; Fernández-Real, J.-M.; et al. Impact of COVID-19 Lockdown in Eating Disorders: A Multicentre Collaborative International Study. *Nutrients* **2022**, *14*, 100. [CrossRef] [PubMed]

37. Murray, J.H.; Mabila, S.L.; McQuistan, A.A. Trends in the Incidence of Eating Disorders Among Active Component Service Members, 2017 to 2021. *MSMR* **2023**, *30*, 19–25. Available online: https://www.health.mil/News/Articles/2023/01/01/Incidence-of-Eating-Disorders (accessed on 1 July 2023). [PubMed]
38. Kessler, R.M.; Hutson, P.H.; Herman, B.K.; Potenza, M.N. The neurobiological basis of binge-eating disorder. *Neurosci. Biobehav. Rev.* **2016**, *63*, 223–238. [CrossRef] [PubMed]
39. Giel, K.E.; Bulik, C.M.; Fernandez-Aranda, F.; Hay, P.; Keski-Rahkonen, A.; Schag, K.; Schmidt, U.; Zipfel, S. Binge eating disorder. *Nat. Rev. Dis. Primers* **2022**, *8*, 16. [CrossRef] [PubMed]
40. Mohammadi, M.R.; Mostafavi, S.A.; Hooshyari, Z.; Khaleghi, A.; Ahmadi, N. Body Mass Index Status across Different Psychiatric Disorders in a National Survey amongst Children and Adolescents: To Identify the Role of Gender. *Iran J. Psychiatry* **2019**, *14*, 253–264. [CrossRef] [PubMed]
41. Miskovic-Wheatley, J.; Koreshe, E.; Kim, M.; Simeone, R.; Maguire, S. The impact of the COVID-19 pandemic and associated public health response on people with eating disorder symptomatology: An Australian study. *J. Eat. Disord.* **2022**, *10*, 9. [CrossRef]
42. Monteleone, A.M.; Cascino, G.; Barone, E.; Carfagno, M.; Monteleone, P. COVID-19 pandemic and eating disorders: What can we learn about psychopathology and treatment? A systematic review. *Curr. Psychiatry Rep.* **2021**, *23*, 83. [CrossRef]
43. Machado, P.P.P.; Pinto-Bastos, A.; Ramos, R.; Rodrigues, T.F.; Louro, E.; Gonçalves, S.; Brandão, I.; Vaz, A. Impact of COVID-19 lockdown measures on a cohort of eating disorders patients. *J. Eat. Disord.* **2020**, *8*, 57. [CrossRef]
44. Mulugeta, W.; Desalegn, H.; Solomon, S. Impact of the COVID-19 pandemic lockdown on weight status and factors associated with weight gain among adults in Massachusetts. *Clin. Obes.* **2021**, *11*, e12453. [CrossRef] [PubMed]
45. Yaqoob, A.; Majeed, I.; Khalid, H.; Hussain, S.; Shahid, M.H.; Majeed, F.; Hassan, O.-U.; Haq, I.-U.; Qing, F. Risk factors for eating disorders among Chinese and international university students: A comparative cross-sectional study. *Cent. Eur. J. Public. Health* **2022**, *30*, 241–247. [CrossRef] [PubMed]
46. COVID-19 Mental Disorders Collaborators. Global prevalence and burden of depressive and anxiety disorders in 204 countries and territories in 2020 due to the COVID-19 pandemic. *Lancet* **2021**, *398*, 1700–1712. [CrossRef] [PubMed]
47. Fancourt, D.; Steptoe, A.; Bu, F. Trajectories of anxiety and depressive symptoms during enforced isolation due to COVID-19 in England: A longitudinal observational study. *Lancet Psychiatry* **2021**, *8*, 141–149. [CrossRef] [PubMed]
48. Bosch, O.J.; Young, L.J. Oxytocin and Social Relationships: From Attachment to Bond Disruption. *Behav. Pharmacol. Neuropept. Oxytocin* **2018**, *35*, 97–117. [CrossRef]
49. Iodice, S.; Ceresa, A.; Esposito, C.M.; Mucci, F.; Conti, D.M.; Pergoli, L.; Tarantini, L.; Vigna, L.; Bollati, V.; Buoli, M.; et al. The Independent Role of Body Mass Index (BMI) and Severity of Depressive Symptoms on Biological Changes of Women Affected by Overweight/Obesity. *Int. J. Environ. Res. Public Health* **2021**, *18*, 2923. [CrossRef] [PubMed]
50. Kerem, L.; Lawson, E.A. The Effects of Oxytocin on Appetite Regulation, Food Intake and Metabolism in Humans. *Int. J. Mol. Sci.* **2021**, *22*, 7737. [CrossRef]
51. Chen, C.Y.; Chiang, Y.C.; Kuo, T.-C.; Tam, K.-W.; Loh, E.-W. Effects of intranasal oxytocin in food intake and craving: A meta-analysis of clinical trials. *Clin. Nutr.* **2021**, *40*, 5407–5416. [CrossRef]
52. Haripersad, Y.V.; Kannegiesser-Bailey, M.; Morton, K.; Skeldon, S.; Shipton, N.; Edwards, K.; Newton, R.; Newell, A.; Stevenson, P.G.; Martin, A.C. Outbreak of anorexia nervosa admissions during the COVID-19 pandemic. *Arch. Dis. Child.* **2021**, *106*, e15. [CrossRef]
53. Martínez-González, M.A.; Gual, P.; Lahortiga, F.; Alonso, Y.; de Irala-Estévez, J.; Cervera, S. Parental factors, mass media influences, and the onset of eating disorders in a prospective population-based cohort. *Pediatrics* **2003**, *111*, 315–320. [CrossRef]
54. Ferrer-Garcia, M.; Pla-Sanjuanelo, J.; Dakanalis, A.; Vilalta-Abella, F.; Riva, G.; Fernandez-Aranda, F.; Sánchez, I.; Ribas-Sabaté, J.; Andreu-Gracia, A.; Escandón-Nagel, N.; et al. Eating behavior style predicts craving and anxiety experienced in food-related virtual environments by patients with eating disorders and healthy controls. *Appetite* **2017**, *117*, 284–293. [CrossRef]
55. Hammerle, F.; Huss, M.; Ernst, V.; Bürger, A. Thinking dimensional: Prevalence of DSM-5 early adolescent full syndrome, partial and subthreshold eating disorders in a cross-sectional survey in German schools. *BMJ Open* **2016**, *6*, e010843. [CrossRef]
56. Galmiche, M.; Déchelotte, P.; Lambert, G.; Tavolacci, M.P. Prevalence of eating disorders over the 2000-2018 period: A systematic literature review. *Am. J. Clin. Nutr.* **2011**, *109*, 1402–1413. [CrossRef]

nutrients MDPI

Article

Community Women's Lifestyle and Eating Disorders in the Era of COVID-19 Pandemic: A 15-Year Follow-Up Study

Nasim Foroughi *, Phillipa Hay and Haider Mannan

Translational Health Research Institute, School of Medicine, Western Sydney University, Locked Bag 1797, Penrith, NSW 2751, Australia
* Correspondence: n.foroughi@westernsydney.edu.au; Tel.: +61-(02)-4634-4580; Fax: +61-(02)-4620-3891

Abstract: Most studies suggest that COVID-19 has adversely affected the quality of life and mental health, including eating disorders. However, studies have yet to examine longitudinally the impact of COVID-19 on eating disorder symptomatic individuals. This study aims to examine longitudinally the impact of the COVID-19 pandemic on the lifestyle and eating disorder symptoms of a symptomatic group of community-dwelling women. These women (n = 171) were enrolled in a longitudinal study, completed a COVID-19 modular self-report (post or Qualtrics, 2020/21), and participated in the current study. This study examined a 15th year follow-up. In 2020, 40% were tested for COVID-19. Of these, 87% had negative results; 5.3% self-isolated at home; 20.5% stopped working/studying in person; 28% continued online work/study; and 28% stopped work/studying in person. The pandemic affected sporting activities, music, and club activities (32.7% discontinued); 38% socialized in person; 16% socialized online; and 10% completely stopped socializing. Findings showed that the respondents showed no significant changes in levels of psychological distress (K10: 21.4 ± 9.8 vs. 19.0 ± 7.1, $p < 0.171$), and impaired quality of life (SF12: 50.9 ± 8.0 vs. 48.3 ± 9.5, $p < 0.055$) at 15-year follow-up. Eating disorder symptoms increased over time (EDE-Q global: 2.1 ± 1.4 vs. 2.9 ± 1.4, $p < 0.013$). Observed worsening of eating disorder-related symptoms during the COVID-19 pandemic may be due to interrupted eating patterns, exercise restrictions and the absence of social support. Provision and access to interventions to support those affected by eating disorders are a high priority, especially during these times.

Keywords: eating disorders; COVID-19; mental health; health care; health literacy; eating disorder-related Medicare item

1. Introduction

Although some of the previously published studies present with conflicting evidence, such as reduced self-harm behavior and suicide [1], most studies suggest that the 2019 novel coronavirus (COVID-19) infection has had an overall, substantially negative and significant psychological, financial, economic, and social impact on mental health globally (including anxiety, psychiatric disorders, depression, eating disorders), as well as on overall quality of life [2–6], A survey conducted by Ammar et al. in 2020 revealed that various factors, such as food type, frequency of food consumption, and meal regularity, have been affected by the pandemic [7]. One major impact of COVID-19 on food consumption is a shift in food type. Due to lockdowns and supply chain disruptions, individuals have been forced to alter their dietary choices. As a result, many have resorted to consuming cheaper and more processed foods, which may not provide the necessary nutrients required for optimal health. The increased stress and anxiety brought on by the pandemic have also led to a rise in comfort eating, which often involves consuming unhealthy foods [8]. With the closure of schools and offices, many individuals have been forced to work or study from home, leading to changes in their daily routine. This has resulted in irregular eating patterns, with many individuals skipping meals or consuming them at irregular times.

In addition, the increased availability of snacks and food at home has also led to more frequent snacking and overeating [9]. Lastly, with the pandemic and associated restrictions limiting individuals' ability to engage in physical activity, many have reported decreased appetite and reduced food intake. This has led to concerns about the potential impact on nutritional status and overall health.

Devoe et al. (2023) recently conducted a systematic review of the impact of the COVID-19 pandemic on the physical and mental health of people with eating disorders. Fifty three studies were included in this review, including participants with common eating disorders such as anorexia nervosa, binge eating disorders, bulimia nervosa, and other specified feeding or eating disorders [10]. The results of this review showed that during the COVID-19 pandemic the rate of hospital admission, eating disorder symptoms, depression and anxiety and social isolation increased during the lockdown, as well as body mass index (BMI) [10]. The authors of this systematic review suggested that the reported worsening of eating disorder symptoms in the previous qualitative literature may be explained by the lack of accessibility to treatment and care, disrupted routine and structure, negative impact of social media, and lockdown [10]. The changes reported by the above-mentioned systematic review, however, appear to be specific to age, are diagnostic, and time, for example. improved symptoms after the lockdown period.

It is evident that the stress and uncertainty of COVID-19 has been adversely affecting the prognosis and quality of life (QOL) of those affected by eating disorders. Previous research has suggested that the pandemic may have influenced both the severity of symptoms and the rate of relapse in patients diagnosed with eating disorders [11]. In addition, mass death, increased levels of unemployment, overwhelmed hospitals, and concern about financial and food insecurity contributed to the increased vulnerability to the emotional impact of COVID-19 and the worsening of psychological distress. With restrictions in place, lack of social support and interaction, mandated lockdown and other psychiatric stress factors occurring due to the direct or indirect impact of the coronavirus pandemic, it has been increasingly difficult to cope for those affected by eating disorders. However, these results contrast with the findings of another study by Machado et al. (2020), in which eating disorder symptoms and BMI did not change post-lockdown in patients with anorexia nervosa, binge eating disorders, bulimia nervosa and other specified feeding or eating disorders [12].

In 2020, a nationwide research project with the aim of assessing the coping strategies related to mental health during the COVID-19 pandemic was performed in Australia [6]. This study reported a statistically significant increase in the levels of eating disorders in general, including binge eating disorders, anorexia nervosa and bulimia nervosa. More specifically, meal avoiding, grazing behavior, overeating, loss of control, overeating, and binge eating episodes have been reported during lockdown [13]. There has been interplay between these eating disorders and isolation following infectious disease outbreaks and the subsequent exacerbation of psychological disorders, as well as deterioration in the quality of life. For example, following the SARS (severe acute respiratory syndrome) outbreak in 2003, a study explored the development of psychological disorders following isolation. Approximately one fifth of the participants of this study reported a clear correlation between mental distress following isolation and the exacerbation of restrictive as well as binge eating disorders [14].

The environment created by the novel coronavirus pandemic may similarly contribute to the development of an eating disorder in previously unaffected individuals. This is demonstrated by the fact that there have been studies indicating a strong positive correlation between the Eating Disorders Symptom impact scale (EDSIS) and the General Health Questionnaire (GHQ). Further research needs to be carried out to find or to further establish direct implications of COVID-19 in eating disorders. Understanding these changes can help to inform public health policies and interventions aimed at promoting healthy eating habits during these unprecedented times. Therefore, the main objective of the current study is to further explore longitudinally the effects of the COVID-19 pandemic on mental health related eating disorder symptoms, quality of life, and depression and anxiety in a moderately symptomatic eating disorder sample, more specifically at 15-year follow-up.

Further, we aimed to investigate the awareness and the use of newly introduced eating disorder Medicare item numbers by the Australian government in this population.

2. Materials and Methods

2.1. Study Design

We obtained ethical approval from the Western Sydney University's Human Research Ethics Committee for this observational analytical study (H13832, August 2020). Participants of a research database established 15 years ago [15] (The Women's Eating and Health Literacy (WEHL)), who gave consent to be contacted for future studies, were contacted and recruited for the current project. This sample consisted of a combination of young women living in Australian Capital Territory and a sample of higher education university and college students. The participants were recruited 15 years ago (baseline) and have been followed up at year 2, 4, 9, 12, and 15. Surveys were distributed to the participants and were completed online via Qualtrics, or by post. All the participants were requested to provide informed consent for participation in the study.

2.2. Assessment Instruments

Depression and anxiety symptoms [16] were assessed via the Kessler-10 (K10) questionnaire, which has been reported to have moderate reliability [17]. An average score indicating positive mental health ranges from 10–15 on this scale [18]. Participant's physical and mental health [19] was measured via the SF12 questionnaire. This questionnaire has sound reliability and validity, reported previously [20]. The scoring is from 0–100 (lowest to highest health level) with the mean and standard deviation score of 50 ± 10 [19]. The Eating Disorder Examination Questionnaire (EDE-Q) was used to assess eating disorder symptoms over the past 28 days prior to study participation [21]. Subscale scores and a global score are derived and Australian community normative data for these have been published previously [22]. This scale has acceptable internal consistency, test–retest reliability and temporal stability. In addition, we collected information on the personal views of the participants and their experiences of the COVID-19 pandemic. The participants were asked about the use and accessibility of the specific Eating Disorder Medicare item numbers introduced by the Australian government in recent years. We asked three questions of the participants and the answers were yes or no.

2.3. Statistical Analysis

We used IBM Statistical Package (SPSS version 23) and SAS (version 9.4) for the analysis of the data in this study. The data was inspected for normality. Chi-square and *T* tests were employed for $p \leq 0.05$. The quantitative data were expressed as mean and standard error (SE) or median and interquartile range after being statistically and visually tested for normality. Open-ended responses were collated and presented by frequency of response. As missing data were low (<10%), no imputation was made. A linear mixed model (marginal model) was used to estimate the trends in eating disorder and quality of life outcomes longitudinally over time. A test of linear trend was performed using *t* test for $p < 0.05$. A rejection of this test indicated that there was a linear trend in the effect of time.

3. Results

The response rates of the mailed surveys (n = 231) and emailed surveys (n = 290) to the previously established research database were 26% (n = 51) and 62.4% (n = 181), respectively. The total response rate was 46.3%. The majority of our participants were married (44%), were employed full time (48.3%), and had a higher education or university degree (83.1%). The survey respondents had high levels of psychological distress, and impaired quality of life (compared to normative data published previously), as presented in Table 1 (participant features). As shown in Table 2, most participants reported using at least one form of complementary therapy 12-months prior to study participation to treat their emotional or mental health problems (58.6%), eating disorders (16.8%), or to lose weight (31%). The most widely utilized complementary therapies were "talking about the problem" and "getting fit by doing more exercise". Favorable attitudes towards complementary therapies and experiences of using them are reported in Table 2.

Table 1. Participant's demographic and health related features (n = 232).

Variables	N (%)
Marital status	
Single	34 (14.7)
Married	102 (44.0)
Living as married	47 (20.3)
Separated or divorced	33 (14.2)
Widowed	2 (0.9)
Occupation	
Employed fulltime	112 (48.3)
Employed Part time	51 (22)
Home maker ^	9 (3.9)
Student	6 (2.6)
Not employed, recovering, other	25 (10.8)
Education	
Year 10 & 12	15 (6.5)
Trade certificate	1 (0.4)
Undergraduate	9 (3.9)
University degree	193 (83.1)
	Mean (SD), Median (Range), n
BMI (kg/m^2)	28.2 (6.7), 26.7 (39.01), 211
EDE-Q	
Restraint	2.9 (1.7), 2.8 (7.0), 209
Eating Concern	2.2 (1.4), 1.6 (6.20), 205
Weight concern	3.3 (1.6), 3.0 (6.20), 210
Shape concern	3.8 (1.7), 3.9 (6.25), 208
Global score	3.0 (1.4), 2.7 (6.04), 194
Psychological distress (Kessler 10)	19.1 (7.7), 17.0 (40.0), 208
Physical health related quality of life (SF-12)	48.4 (9.5), 51.7 (46.7), 203
Mental health related quality of life (SF-12)	45.5 (12), 50.4 (48.2), 203

Home maker: ^ a person who is not employed or student, nor looking for a job, and who is attending house duties in their home; SD: Standard Deviation; BMI: Body Mass Index; EDE-Q: Eating Disorder Examination Questionnaire [21], Kessler 10 [17] components of SF12, including physical wellbeing and mental wellbeing [19].

Table 2. Treatments or activities within the past 12 months prior to study (N = 232).

	Emotional Ormental Health Problems	Eating Disorders	Lose Weight
		N (%)	
Used some type of treatment	136 (58.6)	39 (16.8)	72 (31)
Talked about the problem	110 (47.4)	17 (7.3)	11 (4.7)
Psychotherapy (focus on cause)	42 (18.1)	6 (2.6)	4 (1.7)
Psychotherapy (Cognitive Behaviour Therapy)	45 (19.4)	9 (3.9)	9 (3.9)
Psychotherapy (focus on current relationships)	38 (16.4)	3 (1.3)	2 (.9)
Alternative therapy (naturopathy, homeopathy, aromatherapy)	22 (9.5)	2 (.9)	5 (2.2)
Massage, spinal manipulation, or acupuncture	62 (26.7)	1 (.4)	3 (1.3)

Table 2. *Cont.*

	Emotional Ormental Health Problems	Eating Disorders	Lose Weight
Relaxation therapy (meditation, stress management, yoga)	72 (31)	6 (2.6)	7 (3.0)
Assertiveness or social skills training	9 (3.9)	1 (.4)	1 (.4)
Drinking alcohol to relax	60 (25.9)	3 (1.3)	3 (1.3)
Hypnosis	5 (2.2)	2 (.9)	3 (1.3)
Following a self-help treatment manual	25 (10.8)	6 (2.6)	9 (3.9)
Getting fit by increasing time spent on exercise	80 (34.5)	17 (7.3)	57 (24.6)
Getting out and about more or finding some new hobbies	60 (25.9)	5 (2.2)	8 (3.4)
Getting information about the problem and available services	45 (19.4)	5 (2.2)	9 (3.9)
Medication (valium, serapax)	25 (10.8)	3 (1.3)	3 (1.3)
Anti-depressant medication (Prozac or Zoloft)	42 (18.1)	4 (1.7)	4 (1.7)
Vitamins and/or minerals (folate, B group vitamins, amino acids etc)	55 (23.7)	5 2.2	5 (2.2)
Herbal medicines (Valerian, St John's Wort, ginseng)	19 (8.2)	4 1.7	6 (2.6)

According to the results of our study, only 1.7% of the participants had attempted using an eating disorder related Medicare item. Approximately 87.1% of the participants were not aware of the existence of the eating disorder Medicare item or had not attempted to use it. These results are presented in Table 3.

Table 3. Eating disorder related Medicare item access (N = 232).

Medicare Item Use	N (%)
Attempted using the items	2 (0.9)
Accessed & used the items	4 (1.7)
No attempt/not aware of it	202 (87.1)

Table 4 shows the influence of COVID-19 on mental health, eating disorders and quality of life. Less than half of the participants (46.1%) took the COVID-19 test. None of these participants tested positive for COVID-19. The in-person work-related activities of the participants were not affected by COVID-19 (41.8%). However, 30% of participants stopped attending other in-person activities, such as music and sporting events or lessons. The in-person social activities of the participants were affected to some extent (13.4%). COVID-19 did not affect the eating behavior of 51.3% of the participants. Only 5.6% of the participants were extremely worried about contracting the virus. The information delivery concerning COVID-19 by the government was rated quite clear and understandable by 38.8% of the participants.

Table 4. COVID-19's influence on mental health, eating disorders and quality of life (N = 232).

	N (%)
Had COVID-19	0
Tested for COVID-19	107 (46.1)
Self isolated for 14 days	15 (6.5)
CVID-19's effect on Work and/or study	
Stopped attending work and/or study in person	48 (20.7)
Continued work and/or study at home (online)	76 (32.8)

Table 4. *Cont.*

	N (%)
Stopped work and/or study all together	9 (3.9)
Continued work and/or study in person	97 (41.8)
COVID-19's effect on other activities such as sport, music, clubs, etc.	
Stopped attending activities in person	71 (30.6)
Continued activities at home (online)	63 (27.2)
Stopped activities all together	17 (7.3)
Continued activities in person	54 (23.3)
COVID-19's effect on social life	
Stopped socializing in person	31 (13.4)
Socializing via phone (online)	45 (19.4)
Stopped socializing all together (even online)	26 (11.2)
Continued socializing in person	87 (37.5)
COVID-19's effect on household such as family and housemates	
One or more persons had COVID-19	1 (0.4)
One or more persons were unable to work	14 (6.0)
One or more persons are essential health worker	15 (5.2)
I am an essential health worker	62 (26.7)
None of the above	129 (55.6)
COVID-19's effect on eating	
Not at all	119 (51.3)
More disordered eating (overeating, binge)	59 (25.4)
More dieting and/or exercise for weight control/loss	29 (12.5)
More weight/shape/eating concerns	39 (16.8)
Weeks unbale to attend work	
Not applicable	107 (46.1)
Unable to attend	98 (42.2)
Worried about contracting COVID-19	
Not at all	56 (24.1)
A little worried	81 (34.9)
Moderately worried	49 (21.1)
Very worried	11 (4.7)
Extremely worried	13 (5.6)
Worried about a family member contracting COVID-19	
Not at all	27 (11.6)
A little worried	79 (34.1)
Moderately worried	48 (20.7)
Very worried	39 (16.8)
Extremely worried	17 (7.3)
Rate the information delivery about COVID-19	
Very clear and understandable	35 (15.1)
Quite clear and understandable	90 (38.8)
Neutral	37 (15.9)
Quiet confusing	30 (12.9)
Very confusing	16 (6.9)

Preliminary findings of the current study showed that the respondents had no significant changes in the levels of psychological distress measured via K10 (K10: 21.4 $\pm$ 9.8 vs. 19.0 $\pm$ 7.1, $p < 0.171$), and impaired quality of life (SF12: 50.9 $\pm$ 8.0 vs. 48.3 $\pm$ 9.5, $p < 0.055$) at 15-year follow-up time point. Eating disorder symptoms increased over time (EDE-Q global: 2.1 $\pm$ 1.4 vs. 2.9 $\pm$ 1.4, $p < 0.013$). Eating disorder symptom severity, including restraint, weight concern, shape concern and eating concerns, had all increased from baseline and reached their peak at year 15 when the COVID-19 pandemic was at its highest ($p < 0.0001$ for all). Psychological distress scores increased gradually from baseline to year 9 (2014), had a decline from year 9 to year 12 and reached their highest score after year 12 ($p < 0.0001$). Physical health-related quality of life decreased over time until year 12, after which there was an increase at year 15 ($p < 0.0001$). There was a similar trend for mental health related quality of life; however, the changes were smaller and non-significant ($p = 0.158$). See Figures 1–6 below.

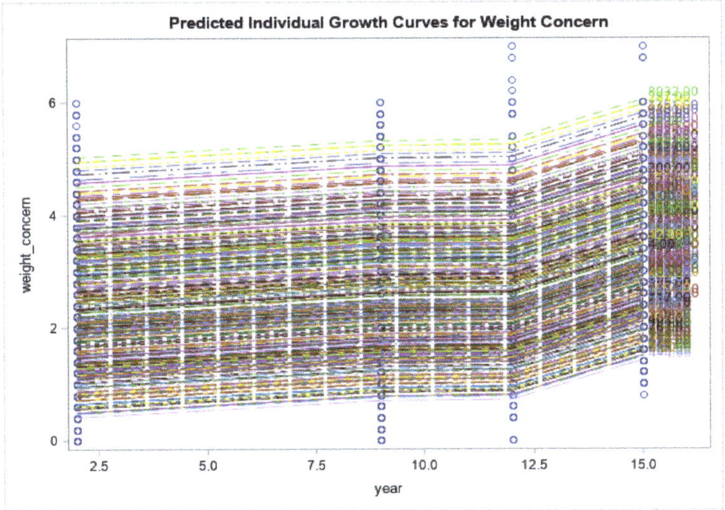

Figure 1. Predicted individual growth curves for weight concern. The blue lines indicate the measurement time points.

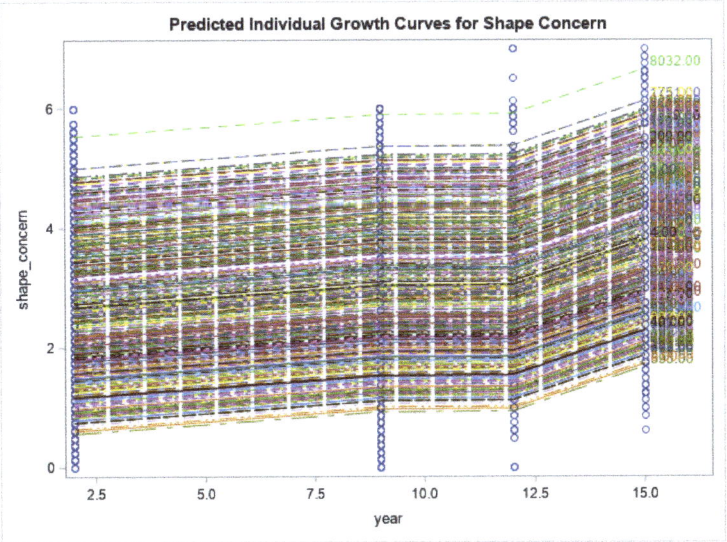

Figure 2. Predicted individual growth curves for shape concern. The blue lines indicate the measurement time points.

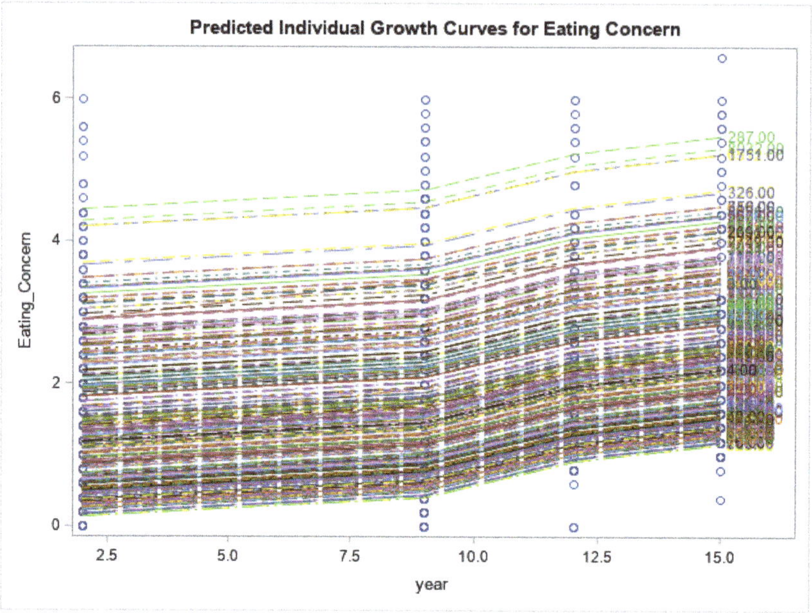

Figure 3. Predicted individual growth curves for eating concern. The blue lines indicate the measurement time points.

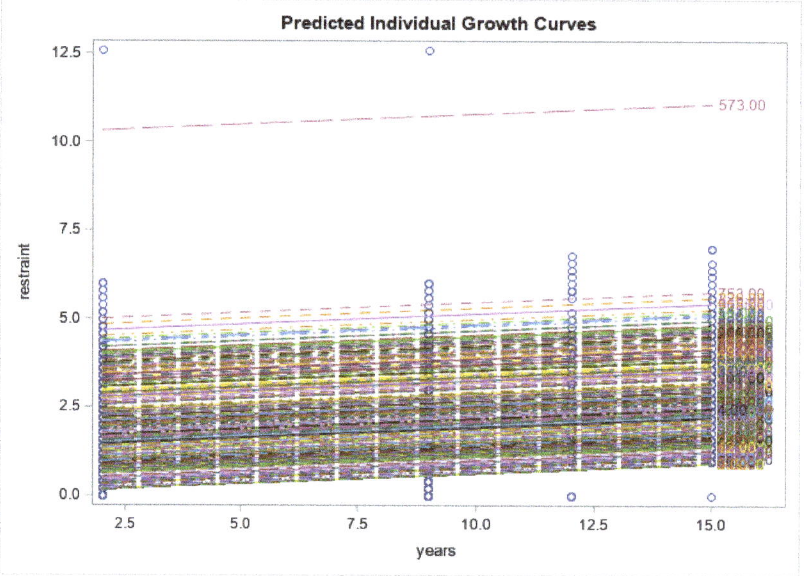

Figure 4. Predicted individual growth curves for eating restraint. The blue lines indicate the measurement time points.

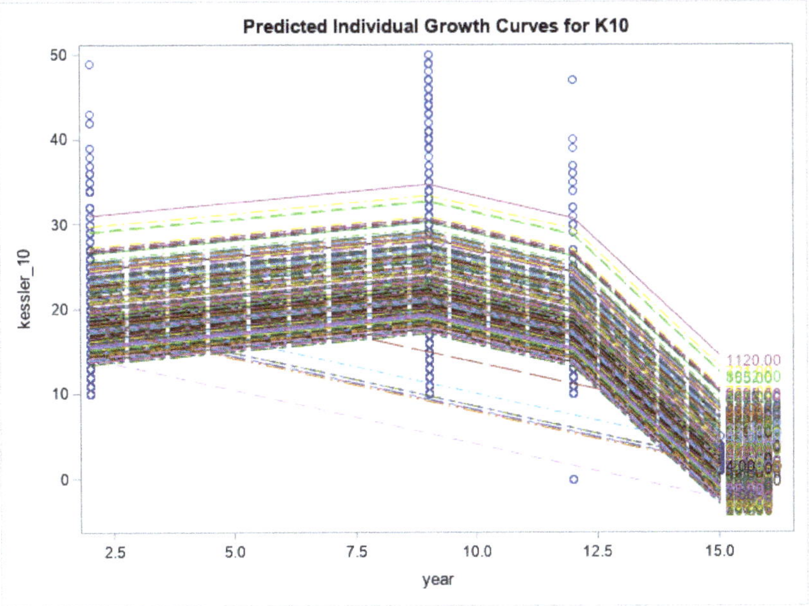

Figure 5. Predicted individual growth curves for K10. The blue lines indicate the measurement time points.

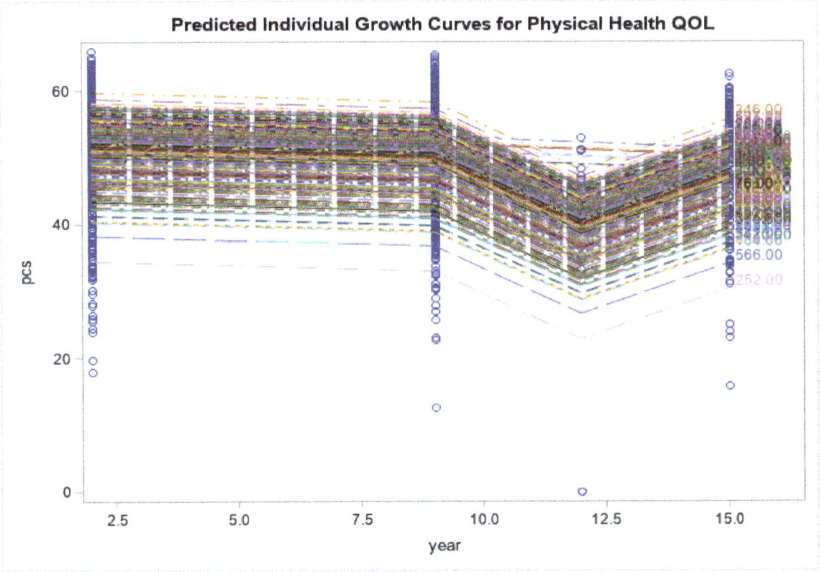

Figure 6. Predicted individual growth curves for physical health related. The blue lines indicate the measurement time points.

4. Discussion

In this current study of moderately symptomatic participants, we examined the effects of the COVID-19 pandemic on eating disorder symptoms and health related quality of life over 15 years in a symptomatic group of community dwelling participants. Overall, eating disorder symptom severity including restraint, weight concern, shape concern and eating concern increased since baseline at year one and reached its peak at year 15 (2020). This was noted when the COVID-19 pandemic was at its highest. In this study, we observed a decline in psychological distress score of the participants from year 9 to year 12. However, these scores then increased over time and reached their peak at year 15 follow-up. Physical health related quality of life decreased with time from baseline to year 12 and then increased again from year 12 to year 15. Mental health related quality of life followed the same trend; however, the changes were not significant.

The results of our study align with previous research conducted on the general population without any history of eating disorders, which has reported a significant decline in mental health during the COVID-19 pandemic. This decline was evident through the self-reported worsening of various symptoms, such as anxiety, sadness, anger, loneliness, and sleep problems, among others. [23–28]. However, some studies have also reported a slight decrease in the mental health problems faced by the general population after the ease of restrictions, compared to the initial negative impacts of the COVID-19 pandemic [29]. Despite these improvements, the changes did not reach the pre-pandemic levels of mental wellbeing [30]. Nonetheless, it is important to note that, despite the improvements, the current state of mental health among the general population still remains a matter of concern.

Similarly, eating disorder symptom severity increased post-pandemic, particularly in patients with eating disorders [31]. Further, Ramalho et al. (2022) found that psychological distress, measured by K10 questionnaire, mediated disordered eating behaviors in a cross-sectional study of adults during the first COVID-19 lockdown period [13]. However, there are some inconsistent findings, particularly in a clinical population [24], where it is possible that some people experience positive effects of confinement due to less exposure to societal judgment and criticism (e.g., from family, friends, and doctors). In addition, more recent longitudinal studies found a rapid decline in distress levels of patients with eating disorders soon after the first increase post-pandemic [32]. On the other hand, a recent systematic review by Devoe et al. (2023) showed increased hospital admission, increased eating disorder symptoms, anxiety and depression during the pandemic lockdown [10]. However, most studies included in this review failed to mention the effects of possible confounders and adjust for important covariates in their analysis, for example, there was a lack of adjustment based on gender or age. This emphasizes the importance of further examining the temporal changes in mental health and eating disorder symptoms of patients with eating disorders in the future.

The findings of the current study showed that the COVID-19 lockdown imposed significant changes on eating behaviors, physical health related quality of life and psychological distress. However, mental health related quality of life remained unchanged. This may be because our study is underpowered, or may reflect the complexity of the construct of mental health related quality of life. Mental health related quality of life measures the impact of poorer mental health on a person's perceived quality of life, but this impact may be reduced where external factors are operating, and mental health related quality of life may be less sensitive to change than a specific measure of psychological distress. During the COVID-19 pandemic, there may have been a positive effect on quality of life because of reduced social interactions and social activities around food. For people with eating disorders, such social interactions may trigger social anxiety [33]. Nevertheless, our results emphasize the importance of designing interventions to alleviate the consequences of the COVID-19 pandemic on eating behaviors or eating disorders, creating a basis for upcoming situations involving lockdown and social isolation.

With regards to the question as to whether the COVID-19 pandemic may be more associated with increased severity and prevalence of eating disorders than with new cases (or incidence), we observed a peaking and rising trajectory in eating disorder symptoms. This was evident after the onset of the COVID-19 pandemic in both people with low, as well as those with high, pre-COVID-19 eating disorder symptom severity [34,35]. Notably, this was not due to increased availability of eating disorder services, as very few participants were aware of the existence of these services (under 3%).

Strengths and Limitations

This longitudinal study aimed to investigate the effects of the COVID-19 pandemic on an eating disorder symptomatic cohort. Most notably, this study is also the first to report on public awareness of and use of the eating disorder related Medicare item in Australia. This study had some important limitations to consider when interpreting the results, including the use of self-reported surveys. These may be prone to reporting and recall biases. For the purpose of this study, we recruited a combination from the general population [22,36]. This cohort was representative of young woman who were living in the Australian Capital Territory, and was a convenience sample of higher education university and college students. The population was broadly representative of young women with at least a high school education in Australia [37–39]. The surveys were distributed to campuses with diverse socio-economic backgrounds. Although there is a very high proportion (56%) of the Australian population who engage in higher education [40], this sample was not representative of the wider Australian population. Therefore, the results may not be generalizable to other genders than females, to senior citizens, and to people with no eating disorder symptoms. The findings of the current study may provide useful information for future studies to examine, to a greater extent, issues concerning mental health, for example, differences in gender, age, and education. Such groups may be imported to target interventions aimed at improving mental health in the future.

5. Conclusions

The COVID-19 crisis has presented a number of challenges for individuals struggling with eating disorders. These challenges have the potential to worsen the symptoms of the disorder and increase the risk of developing or worsening disordered eating behaviors. For women in particular, the pandemic has exacerbated psychological distress and negatively impacted their physical health-related quality of life. The pandemic has resulted in significant changes to daily life for many people. Isolation, job loss, financial stress, and uncertainty about the future have all contributed to an increase in anxiety and depression. For individuals with eating disorders, these factors can trigger or worsen their symptoms. The disruption to daily routines, changes in food availability, and reduced access to healthcare services have also added to the challenge of managing eating disorders.

Women are particularly vulnerable to the negative effects of the pandemic on mental and physical health. Women are more likely to experience economic hardship, work in industries with high levels of exposure to the virus, and take on additional caregiving responsibilities. These stressors can further exacerbate the psychological distress associated with eating disorders and increase the risk of developing new symptoms. Overall, the COVID-19 crisis has highlighted the need for increased support and resources for individuals struggling with eating disorders, particularly for women, who may be disproportionately impacted. It is important for healthcare providers to be aware of the unique challenges faced by individuals with eating disorders during this time and to provide appropriate care and support.

Increasing community awareness of the existence and availability of the eating disorder Medicare item newly introduced by the Australian government may be beneficial for those affected by eating disorders in general and specifically during a crisis such as the COVID-19 pandemic. It is crucial for people with eating disorders to obtain support and have access to resources, even during these challenging times. Provision and access to interventions to support the people affected by eating disorders are a high priority, especially during these times and similar future situations. Overall, further research is needed to explore the impacts of COVID-19 and other pandemics on individuals with eating disorders. Such research should aim to identify effective strategies to reduce the negative impacts of pandemics on eating disorder patients and to improve their overall quality of life. For instance, research can explore how telehealth and online resources can be used to provide effective treatment and support to individuals with eating disorders during pandemics. Additionally, researchers can study the long-term impact of the pandemic on the development and course of eating disorders and identify ways to mitigate these effects. Furthermore, research can investigate how to address the unique challenges faced by marginalized communities, such as individuals with low income, people of color, and LGBTQ+ individuals, who may be disproportionately affected by the pandemic and its impact on eating disorders.

Author Contributions: P.H. and N.F. conceived of the presented idea and theory development. N.F. performed the analytical computations and H.M. verified the analytical methods. All authors discussed the results and contributed to the final manuscript. All authors have read and agreed to the published version of the manuscript.

Funding: P.H. receives sessional fees and lecture fees from the Australian Medical Council, Therapeutic Guidelines publication, and New South Wales Institute of Psychiatry and royalties from Hogrefe and Huber, McGraw Hill Education, and Blackwell Scientific Publications, and she has received research grants from the NHMRC and ARC.

Institutional Review Board Statement: The study was conducted in accordance with the Declaration of Helsinki, and approved by the Institutional Review Board (or Ethics Committee) of The Western Sydney University (H13832, August 2020).

Informed Consent Statement: Human Research Ethics Committee of the Western Sydney University has granted approval for this study (Reference H6644). Informed consent has been signed by all the participants.

Data Availability Statement: The data presented in this study are available on request from the corresponding author. The data are not publicly available due to patients' privacy and ethical permissions.

Acknowledgments: The authors acknowledge the participants for their valuable time and effort.

Conflicts of Interest: The authors declare no conflict of interest.

References

1. Dragovic, M.; Pascu, V.; Hall, T.; Ingram, J.; Waters, F. Emergency department mental health presentations before and during the COVID-19 outbreak in Western Australia. *Australas. Psychiatry* **2020**, *28*, 627–631. [CrossRef]
2. Fisher, J.R.; Tran, T.D.; Hammarberg, K.; Sastry, J.; Nguyen, H.; Rowe, H.; Popplestone, S.; Stocker, R.; Stubber, C.; Kirkman, M. Mental health of people in Australia in the first month of COVID -19 restrictions: A national survey. *Med. J. Aust.* **2020**, *213*, 458–464. [CrossRef]
3. Asmundson, G.; Taylor, S. Coronaphobia: Fear and the 2019-nCoV outbreak. *J. Anxiety Disord.* **2020**, *70*, 102–196. [CrossRef] [PubMed]
4. Lima, C.; Carvalho, P.; Lima, I.; Nunes, J.; Saraiva, J.; de Souza, R.; da Silva, C.; Neto, M. The emotional impact of Coronavirus 2019-nCoV (new Coronavirus disease). *Psychiatry Res.* **2020**, *287*, 112–915. [CrossRef]
5. Maunder, R. Was SARS a mental health catastrophe? *Gen. Hosp. Psychiatry* **2009**, *31*, 316–317. [CrossRef] [PubMed]
6. Phillipou, A.; Meyer, D.; Neill, E.; Tan, E.; Toh, W.; Van Rheenen, T.; Rossell, S. Eating and exercise behaviors in eating disorders and the general population during the COVID -19 pandemic in Australia: Initial results from the COLLATE project. *Int. J. Eat. Disord.* **2020**, *53*, 1158–1165. [CrossRef]
7. Ammar, A.; Brach, M.; Trabelsi, K.; Chtourou, H.; Boukhris, O.; Masmoudi, L.; Bouaziz, B.; Bentlage, E.; How, D.; Ahmed, M.; et al. Effects of COVID-19 Home Confinement on Eating Behaviour and Physical Activity: Results of the ECLB-COVID19 International Online Survey. *Nutrients* **2020**, *12*, 1583. [CrossRef]
8. Di Renzo, L.; Gualtieri, P.; Cinelli, G.; Bigioni, G.; Soldati, L.; Attinà, A.; Bianco, F.; Caparello, G.; Camodeca, V.; Carrano, E.; et al. Psychological Aspects and Eating Habits during COVID-19 Home Confinement: Results of EHLC-COVID-19 Italian Online Survey. *Utrients* **2020**, *12*, 2152. [CrossRef]
9. Shahrokni, R.O.; Ferriday, D.; Miguel, S.; Laurent, A.L.; Brunstrom, J.M. Portion size reduction and food choice: The role of perceptual boundaries. *Appetite* **2021**, *157*, 104941. [CrossRef]
10. Husain, M.M.; Black, K.J.; Doraiswamy, P.M.; Shah, S.A.; Rockwell, W.K.; Ellinwood, E.H., Jr.; Krishnan, K.R. Subcortical brain anatomy in anorexia and bulimia. *Biol. Psychiatry* **1992**, *31*, 735–738. [CrossRef]
11. Touyz, S.; Lacey, H.; Hay, P. Eating disorders in the time of COVID-19. *J. Eat. Disord.* **2020**, *8*, 19. [CrossRef] [PubMed]
12. Machado, P.P.P.; Pinto-Bastos, A.; Ramos, R.; Rodrigues, T.F.; Louro, E.; Gonçalves, S.; Brandão, I.; Vaz, A. Impact of COVID-19 lockdown measures on a cohort of eating disorders patients. *J. Eat. Disord.* **2020**, *8*, 57. [CrossRef]

13. Ramalho, S.M.; Trovisqueira, A.; de Lourdes, M.; Gonçalves, S.; Ribeiro, I.; Vaz, A.R.; Machado, P.P.P.; Conceição, E. The impact of COVID-19 lockdown on disordered eating behaviors: The mediation role of psychological distress. *Eat. Weight Disord.* **2022**, *27*, 179–188. [CrossRef]
14. Mihashi, M.; Otsubo, Y.; Yinjuan, X.; Nagatomi, K.; Hoshiko, M.; Ishitake, T. Predictive factors of psychological disorder development during recovery following SARS outbreak. *Health Psychol.* **2009**, *28*, 91–100. [CrossRef] [PubMed]
15. Foroughi, N.; Zhu, K.C.Y.; Smith, C.; Hay, P. The perceived therapeutic benefits of complementary medicine in eating disorders. *Complement. Ther. Med.* **2019**, *4*, 176–180. [CrossRef]
16. Kessler, R.C.; Barker, P.R.; Colpe, L.J.; Epstein, J.F.; Gfroerer, J.C.; Hiripi, E. The Kessler Psychological Distress Scale (K10). Available online: https://www.worksafe.qld.gov.au/__data/assets/pdf_file/0010/22240/kessler-psychological-distress-scale-k101.pdf (accessed on 23 March 2023).
17. Dal Grande, E.; Taylor, A.; Wilson, D. *South Australian Health and Wellbeing Survey*; South Australian Department of Human Services: Adelaide, Australia, 2022.
18. Kessler, R.; Barker, P.; Colpe, L.; Epstein, J.; Gfroerer, J.; Hiripi, E.; Howes, M.J.; Normand, S.L.; Manderscheid, R.W.; Walters, E.E.; et al. Screening for serious mental illness in the general population. *Arch. Gen. Psychiatry* **2003**, *60*, 184–189. [CrossRef]
19. Ware, J.E.; Kosinski, M.; Keller, S.D. A 12-Item Short-Form Health Survey: Construction of scales and preliminary tests of reliability and validity. *Med. Care* **1996**, *34*, 220–233. [CrossRef]
20. Salyers, M.P.; Bosworth, H.B.; Swanson, J.W.; Lamb-Pagone, J.; Osher, F.C. Reliability and validity of the SF-12 health survey among people with severe mental illness. *Med. Care* **2000**, *38*, 1141–1150. [CrossRef] [PubMed]
21. Beglin, S.J.; Fairburn, C.G. disorders in community samples. *Psychiatry Res.* **1992**, *44*, 191–201. [CrossRef]
22. Mond, J.M.; Hay, P.J.; Rodgers, B.; Owen, C. Eating Disorder examination questionnaire (EDE-Q): Norms for young adult women. *Behav. Res. Ther.* **2006**, *44*, 53–62. [CrossRef] [PubMed]
23. Cao, W.; Fang, Z.; Hou, G.; Han, M.; Xu, X.; Dong, J.; Zheng, J. The psychological impact of the COVID-19 epidemic on college students in China. *Psychiatry Res.* **2020**, *287*, 112934. [CrossRef]
24. Fernández-Aranda, F.; Munguía, L.; Mestre-Bach, G.; Steward, T.; Etxandi, M.; Baenas, I.; Granero, R.; Sánchez, I.; Ortega, E.; Andreu, A.; et al. COVID Isolation Eating Scale (CIES): Analysis of the impact of confinement in eating disorders and obesity-A collaborative international study. *Eur. Eat. Disord. Rev.* **2020**, *28*, 871–883. [CrossRef]
25. Kavoor, A.R. COVID-19 in people with mental illness: Challenges and vulnerabilities. *Asian J. Psychiatry* **2020**, *51*, 102051. [CrossRef] [PubMed]
26. Nguyen, H.C.; Nguyen, M.H.; Do, B.N.; Tran, C.Q.; Nguyen, T.T.; Pham, K.M.; Pham, L.V.; Tran, K.V.; Duong, T.T.; Tran, T.V.; et al. People with suspected COVID-19 symptoms were more likely depressed and had lower health-related quality of life: The potential benefit of health literacy. *J. Clin. Med.* **2020**, *9*, 965. [CrossRef] [PubMed]
27. Qiu, J.; Shen, B.; Zhao, M.; Wang, Z.; Xie, B.; Xu, Y. A nationwide survey of psychological distress among Chinese people in the COVID-19 epidemic: Implications and policy recommendations. *Gen. Psychiatry* **2020**, *33*, e100213. [CrossRef]
28. Wang, Y.; Di, Y.; Ye, J.; Wei, W. Study on the public psychological states and its related factors during the outbreak of coronavirus disease 2019 (COVID-19) in some regions of China. *Health Med.* **2020**, *30*, 13–22. [CrossRef]
29. Xiong, J.; Lipsitz, O.; Nasri, F.; Lui, L.; Gill, H.; Phan, L.; Chen-Li, D.; Iacobucci, M.; Ho, R.; Majeed, A.; et al. Impact of COVID-19 pandemic on mental health in the general population: A systematic review. *J. Affect. Disord.* **2020**, *277*, 55–64. [CrossRef] [PubMed]
30. Richter, D.; Riedel-Heller, S.; Zürcher, S.J. Mental health problems in the general population during and after the first lockdown phase due to the SARS-cov-2 pandemic: Rapid review of multi-wave studies. *Epidemiol. Psychiatr. Sci.* **2021**, *30*, e27. [CrossRef] [PubMed]
31. Castellini, G.; Cassioli, E.; Rossi, E.; Innocenti, M.; Gironi, V.; Sanfilippo, G.; Felciai, F.; Monteleone, A.M.; Ricca, V. The impact of COVID-19 epidemic on eating disorders: A longitudinal observation of pre versus post psychopathological features in a sample of patients with eating disorders and a group of healthy controls. *Int. J. Eat. Disord.* **2020**, *53*, 1855–1862. [CrossRef]
32. Daly, M.; Robinson, R. Longitudinal changes in psychological distress in the UK from 2019 to September 2020 during the COVID-19 pandemic: Evidence from a large nationally representative study. *Psychiatry Res.* **2021**, *300*, 113920. [CrossRef]
33. Levinson, C.A.; Brosof, L.C.; Vanzhula, I.; Christian, C.; Jones, P.; Rodebaugh, T.L.; Langer, J.K.; White, E.K.; Warren, C.; Weeks, J.W.; et al. Social anxiety and eating disorder comorbidity and underlying vulnerabilities: Using network analysis to conceptualize comorbidity. *Int. J. Eat. Disord.* **2018**, *51*, 693–709. [CrossRef]
34. Linardon, J.; Messer, M.; Rodgers, R.F.; Fuller-Tyszkiewicz, M. A systematic scoping review of research on COVID-19 impacts on eating disorders: A critical appraisal of the evidence and recommendations for the field. *Int. J. Eat. Disord.* **2022**, *55*, 3–8. [CrossRef] [PubMed]
35. Eddy, K.T.; Tabri, N.; Thomas, J.J.; Murray, H.B.; Keshaviah, A.; Hastings, E.; Edkins, K.; Krishna, M.; Herzog, D.B.; Keel, P.K.; et al. Recovery from anorexia nervosa and bulimia nervosa at 22-year follow-up. *J. Clin. Psychiatry* **2017**, *22*, 17085. [CrossRef]
36. Mond, J.M.; Hay, P.; Rodgers, B.; Owen, C.; Crosby, R.; Mitchell, J.E. Use of extreme weight control behaviors with and without binge eating in a community sample of women: Implications for the classification of bulimic-type eating disorders. *Int. J. Eat. Disord.* **2006**, *39*, 294–302. [CrossRef] [PubMed]
37. Mond, J.; Hay, P.; Paxton, S.; Rodgers, B.; Darby, A.; Nillson, J.; Quirk, F.; Owen, C. Eating Disorders "Mental Health Literacy" in Low Risk, High Risk and Symptomatic Women: Implications for Health Promotion Programs. *Eat. Disord.* **2010**, *18*, 267–285. [CrossRef] [PubMed]

38. Mond, J.M.; Hay, P.J.; Rodgers, B.; Owen, C. Health service utilization for eating disorders: Findings from a community-based study. *Int. J. Eat. Disord.* **2007**, *40*, 399–408. [CrossRef]
39. Mitchison, D.; Morin, A.; Mond, J.; Slewa-Younan, S.; Hay, P. The bidirectional relationship between quality of life and eating disorder symptoms: A 9-year community-based study of Australian women. *PLoS ONE* **2015**, *26*, e0120591. [CrossRef]
40. Australian Bureau of Statistics. Australians Pursuing Higher Education in Record Numbers. 2017. Available online: https://www.abs.gov.au/AUSSTATS/abs@.nsf/mediareleasesbyReleaseDate/1533FE5A8541D66CCA2581BF00362D1D (accessed on 17 September 2021).

nutrients

MDPI

Article

Virtual Rejection and Overinclusion in Eating Disorders: An Experimental Investigation of the Impact on Emotions, Stress Perception, and Food Attitudes

Paolo Meneguzzo [1,2,*], Valentina Meregalli [1,2], Enrico Collantoni [1,2], Valentina Cardi [3,4], Elena Tenconi [1,2] and Angela Favaro [1,2]

1 Department of Neuroscience, University of Padova, Via Giustiniani 2, 35128 Padova, Italy
2 Neuroscience Padova Center, University of Padova, Via Orus 2, 35128 Padova, Italy
3 Department of Psychological Medicine, Institute of Psychiatry, Psychology and Neuroscience, King's College London, London WC2R 2LS, UK
4 Department of General Psychology, University of Padova, Via Venezia 8, 35128 Padova, Italy
* Correspondence: paolo.meneguzzo@unipd.it; Tel.: +39-0498213830

Abstract: (1) Background: the investigation of how interpersonal functioning affects eating psychopathology has been receiving increasing attention in the last decade. This study evaluates the impact of virtual social inclusion or ostracism on emotions, perceived stress, eating psychopathology, and the drive to binge or restrict in patients across the eating disorder spectrum. (2) Methods: a group of 122 adolescent and adult females with different eating disorder diagnoses were compared to 50 healthy peers with regards to their performance on, and responses to the Cyberball task, a virtual ball-tossing game. Each participant was randomly assigned to playing a social inclusion or a social exclusion block of the Cyberball task and completed self-report assessments of emotions, perceived stress and urge to restrict/binge before and after the task. (3) Results: patients with anorexia nervosa showed a more negative impact on psychological well-being evaluated with the need threat scale after the excluding block, while patients with bulimia nervosa reported more negative effects after the overincluding condition. Patients with binge eating disorder showed a reduction in specific negative emotions after the overincluding block, unlike all other participants. (4) Conclusions: findings show significant correlations between restraint thoughts in patients with bulimia nervosa and binge thoughts in patients with binge eating disorder after being exposed to the inclusion condition. Different reactions in cognitive and emotional states of patients with eating disorders after different interpersonal scenarios confirm the impact of inclusive or exclusive relationships on eating psychopathology, with specific and different responses across the eating disorder spectrum, that have been discussed, linked to their eating behavioral cognition.

Keywords: anorexia nervosa; bulimia nervosa; binge eating disorder; Cyberball; ostracism; overinclusion; food attitudes

1. Introduction

Despite differences in their clinical manifestations, people with eating disorders (EDs) often experience interpersonal problems and difficulties with social cognition that negatively affect their social functioning, quality of life, and daily activities [1–5]. Different psychological elements have been related to dysfunctional eating behaviors and interpersonal difficulties [6]. Individuals with EDs have generally reported the presence of maladaptive coping strategies in various social contexts, showing dysfunctional emotional responses that are modulated through eating disorder symptoms, such as caloric restriction, bingeing, purging, or compulsive exercise [6–8]. For example, the fear of negative evaluation and social anxiety are vulnerabilities in individuals with EDs whose symptomatology includes body dissatisfaction, bulimic symptoms, shape concerns, weight concerns, and eating concerns [9,10]. Instead, less is known about the fear of positive evaluation in EDs, even if it is a relevant factor in social anxiety [11,12]. Indeed, the fear of others' positive judgments has been positively correlated to high levels of social and meal-related anxiety, with possible connections to others' expectation, clinical perfectionism, and concerns about mistakes [9,13]. These psychological aspects have been related to dysfunctional behaviors in EDs and are predictors of outcomes [14]. Looking at the general population, fear of positive evaluation has been positively related to bulimic symptoms, body dissatisfaction, weight, and shape concerns [9]. However, data are still preliminary, and no structured study is available about this specific social features in these clinical populations.

Looking at the cognitive aspects of emotional functioning, patients with anorexia nervosa (AN) demonstrate reduced emotional expression, difficulties understanding their own and others' emotional and mental states, submissive attitudes and behaviors, and social inhibition in their relationships [15–18]. Similarly, patients with bulimia nervosa (BN) and binge eating disorder (BED) are struggling with identifying their feelings and managing interpersonal relationships [19]. Although more evidence is needed, these findings suggest a possible impairment in social skills across the entire ED spectrum [3,20–23].

The literature has shown that greater sensitivity to social rejection may lead to social difficulties, along with a more negative interpretation of ambiguous social situations [24]. Since the perception of being excluded or ignored by others represents a potential threat to the fundamental psychological need to be accepted, it is usually experienced as a painful situation and could have negative consequences, including the exacerbation of ED symptoms [25]. Some studies on healthy individuals have shown that being ostracized impairs participants' self-regulation and increases their weight concerns and motivation to eat [26,27]. Looking at clinical samples, one study [28] observed that patients with BN reported an increase in hunger and desire to binge after completing an imagery task designed to elicit feelings of lack of appreciation and rejection. Only a few studies, however, have examined how patients with EDs interpret and react to social interactions, largely focusing on social exclusion [6], with no study available on the effects of social overinclusion. The fear of positive evaluations has been related to eating psychopathology as negative judgments [9,13], but actual overinclusion may be a different construct—and it has been neglected in the literature. Positive evaluations may trigger similar psychopathological responses as negative ones because individuals may interpret them as equally judgmental [29,30]. Yet, the previous literature—perhaps because patients tend toward social isolation—has neglected this aspect of eating psychopathology, and comparisons between social exclusion and inclusion are needed.

One task commonly used to assess participants' reactions to interpersonal dynamics is the Cyberball task [31], a computerized ball-tossing game that can manipulate the degree of social inclusion by varying the number of times a participant is passed the ball. This task has been used to assess exclusion and overinclusion in both healthy individuals and patients with neurodevelopmental and psychiatric conditions such as autism spectrum disorder, major depression, social anxiety, and particularly in borderline personality disorder [32–34]. The Cyberball task has also been applied in studies of interpersonal relationships and eating behaviors to show the effects of social exclusion on food assumptions [35–37]. The Cyberball conditions of exclusion and inclusion/overinclusion have been related several times, respectively, to the fear of negative and positive evaluations [38–40], but studies specifically addressing the ED population are lacking. In a previous study, we implemented the Cyberball task in a sample of patients with AN using exclusion and overinclusion paradigms; our results showed that although patients with AN tended to overestimate their actual exclusion and displayed higher sensitivity to exclusion than their peers, they had no self-reported changes in the mood [41]. Similarly, in the overincluding task, they did not report any difference compared to the controls, corroborating the idea that fear of positive evaluation may be more linked to food anxiety and other social aspects [13], but studies on the population with bulimic symptoms are needed.

Despite the finding that patients with AN experience greater sensitivity to exclusion than healthy peers, it is largely unknown whether people with a different ED diagnosis (i.e., BN or BED) would respond similarly or whether virtually induced social exclusion or overinclusion have a different impact on their emotions, stress levels, and eating attitudes. Thus, the present study seeks to extend previous findings among patients with AN by including participants with BN or BED in our assessment of the impact of virtual exclusion and overinclusion, looking at mood and eating behaviors. Our first aim was to evaluate the effects of exclusion and overinclusion in patients with an ED compared to community peers. We hypothesized that patients with different EDs would present different emotional and cognitive responses to exclusion and overinclusion. Individuals with AN may confirm their lack of response to exclusion, while individuals with BN or BED might be more sensitive to overinclusion, due to the preliminary data in the general population [9]. Our second aim was to evaluate the effects of interpersonal interactions on emotions, stress, and attitudes toward food, examining specific differences between participants who might have relevant clinical effects. Regarding clinical features, more restrictive responses were expected in patients with AN, while more eating-mediated responses were expected in patients with binge episodes. Finally, an exploration of the specific correlations between changes in emotions and attitudes toward food would help clinicians to understand any links between eating behaviors and interpersonal difficulties.

2. Materials and Methods

2.1. Participants

Participants were recruited from the University Hospital of Padova (Italy) at the local Eating Disorder Center. A group of 122 patients (AN ($n = 42$), BN ($n = 40$), and BED ($n = 40$)) was included. For comparison, a group of 50 healthy women (HW) was also recruited from the general population through non-specific public announcements seeking volunteers and during meetings in public high schools and at the University of Padova. Parts of the AN (30 out of 42) and HW (34 out of 50) groups were also included in our previous paper [41]. The inclusion criteria for both groups were as follows: female gender; between 15 and 40 years old; no severe psychiatric (e.g., bipolar disorder or schizophrenia) or medical comorbidity (e.g., severe hypotension, bradycardia, or electrolyte alterations), neurological trauma, or disorder. Additional criteria for the HW group were no lifetime diagnosis of ED, anxiety, or mood disorders. Inclusion and exclusion criteria were evaluated via a clinical interview performed by a trained psychiatrist, and the ED diagnoses agreed with Diagnostic and Statistical Manual of Mental Disorders (DSM 5) criteria [12]. Before participating in the study, each participant (and parents of participants younger than 18 years of age) was given written consent. All participants were volunteers and did not receive any credits or payments. The local ethics committee approved the study as part of a more extensive study on the cognitive evaluation of ED patients, and the study complies with the provisions of the Declaration of Helsinki.

2.2. Materials

Demographic and clinical data (height, weight, age at menarche, education, medications) were directly detected by researchers during the assessment of the participants. This included the duration of the disorder for ED patients, measured by the time (in months) from the beginning of ED symptomatology to the moment of the study.

2.2.1. Self-Reported Measures

The psychological assessment was composed of various self-report questionnaires. The Eating Disorders Examination Questionnaire (EDE-Q), a 28-item questionnaire, assesses eating psychopathology and eating disorder behaviors over the previous 28 days [43]; in this study, the EDE-Q showed an excellent internal consistency ($\alpha = 0.961$). The Patient Health Questionnaire (PHQ-9) is a nine-item self-report questionnaire used to evaluate the severity of depressive psychopathology [44]; in this study, $\alpha = 0.868$. The Positive and

Negative Affect Schedule (PANAS) is a 20-item self-report measure for the evaluation of positive and negative affect (e.g., enthusiasm, activity, fear, shame); items are rated on a five-point Likert scale and clustered into two subscales: PANAS-positive and PANAS-negative [45]; in this study, PANAS-positive $\alpha = 0.877$, PANAS-negative $\alpha = 0.923$. The need threat scale (NTS) is a 20-item questionnaire [31] usually completed after the Cyberball game that assesses feelings of distress on four dimensions: belonging ("I felt rejected"), self-esteem ("I felt liked"), meaningful existence ("I felt invisible"), and control ("I felt powerful") on a five-point scale from "Not at all" to "Extremely", with higher scores indicating greater distress. This scale was specifically designed to evaluate the cognitive and emotional effects of the Cyberball task [46]; in this study, $\alpha = 0.786$. Two more items evaluated on a five-point Likert scale were added to the NTS to assess the extent to which participants felt excluded from the game ("I was ignored" and "I was excluded") and were used to calculate the variable manipulation check, as suggested by the previous literature [47]. The participants were also asked to estimate the percentage of ball tosses they received by marking a point on a 10-cm long visual analogue scale. Finally, they evaluated their own levels of perceived stress (i.e., stress-related scale: SRS) and thoughts about binging or restricting (i.e., thoughts about binge scale: TBS; thoughts about restrict scale: TRS) using a seven-point Likert scale.

2.2.2. Cyberball Task

The Cyberball task is a computerized ball-toss game used to evaluate the psychological effects of exclusion and inclusion in a virtual social situation [31]. Before the task, participants were told they would be playing a virtual ball-toss game with two unknown female participants who were online with them. The software is free—https://www.empirisoft.com/cyberball.aspx (accessed on 1 February 2023)—and the game is performed with a laptop, on which the participants see a graphical representation of each of the other two players near a corner of their screens. The game consisted of either an exclusion or overinclusion block of 30 ball tosses each; participants were randomly allocated to play one block or the other [41,48]. In the exclusion condition, participants received the ball four times (13% of the tosses). In the overinclusion condition, participants received the ball almost every time the computer tossed the ball (46% of the tosses). The ball receiving rates were decided based on previous literature on various psychiatric conditions, where overinclusion was rated as a specific social trigger for participants. In contrast, average inclusion might not be sufficient due to the possible tendency to expect an exclusion in ED population, or the negative interpretation of positive evaluations reported in the previous literature [9,41,49,50].

2.3. Procedure

The study used a single-session design. After signing the consent form, participants completed the self-report questionnaires (EDE-Q, PHQ-9, PANAS, SRS, TBS, and TRS) and were randomly assigned to play either the exclusion or overinclusion block of the Cyberball game. The game was always played in the afternoon and in the same room, using a 17-inch laptop. After playing the game, participants completed the NTS, rated the percentage of ball tosses received, and completed the post-test self-report questionnaires (PANAS, SRS, TBS, and TRS). Finally, participants were debriefed according to the international guidelines regarding deception [51]. See Figure 1 for a graphical representation of the procedure.

2.4. Statistical Analyses

Different analyses of variances (ANOVAs) with Bonferroni-corrected post hoc tests were calculated to examine group differences in sample characteristics, evaluating differences between clinical and non-clinical subgroups separately for each experimental block. A comparison between the participants in the excluded and overincluded conditions was performed for the demographic variables using a *t* test for independent samples. As suggested by the literature [33], a specific variable called manipulation check was created using two specific items included in the NTS scale to explicitly evaluate the participants' perception of being manipulated during the task. The difference between groups playing the exclusion and overinclusion blocks was estimated using a *t* test for independent samples. The pre-and-post Cyberball PANAS, SRS, TBS, and TRS scores were tested with a series of general linear models (GLMs) for repeated measures with Tukey HSD (honestly significant difference) post hoc analysis. A first analysis level was characterized by comparing the two conditions separately, then a series of GLMs with two factors between groups (condition and diagnosis) and one factor within group (time) was performed. Different *t* tests for paired samples were applied for the evaluation of the effect of the Cyberball task and are reported in the Supplementary Materials (Tables S1–S4). Different correlation analyses were performed for each subgroup using Pearson's correlations for the subscales of the NTS and the differences (D, calculated as pre- minus post- scores) in PANAS, SRS, TBS, and TRS scales. The α was corrected for multiple comparisons with the Bonferroni correction, with *p* values equal to or lower than 0.013 considered significant, with the exception of the GLM analysis, which was subject to an HSD correction as specified above. The effect sizes were calculated with partial eta squared (ηp^2). Cronbach's α was measured to evaluate the internal consistency of the questionnaires. All data were analyzed using IBM SPSS Statistics 25.0 software (SPSS, Chicago, IL, USA).

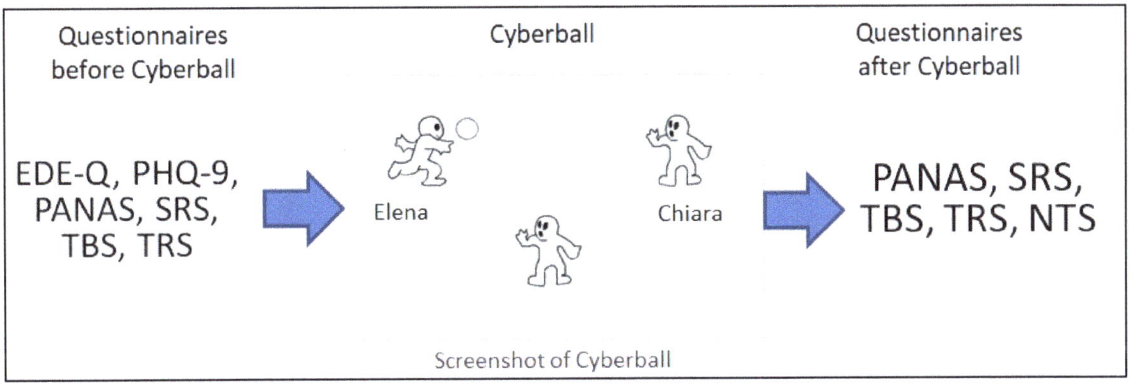

Figure 1. Graphical representation of the procedure. Before the task, participants filled out a series of questionnaires that were partially repeated after the task to record their emotions and thought changes. EDE-Q: eating disorder examination questionnaire; PHQ-9: Patient Health Questionnaire; PANAS: Positive and Negative Affect Schedule; SRS: stress-related scale; TBS: thoughts about binge scale; TRS: thoughts about restrict scale; NTS: need threat scale.

3. Results

The demographic and psychological characteristics of the participants are summarized in Table 1. For the age of participants, underage participants comprised seven with AN (17%), five with BN (13%), three with BED (13%), and five in the HW group (10%), with no significant differences in distribution ($p = 0.201$). Nineteen patients (16%) were being medicated with selective serotonin reuptake inhibitors or benzodiazepines at the time of testing: 10 in the overinclusive condition (four patients with AN, three with BN, and three with BED) and nine in the excluded one (three patients with AN, three with BN, and three with BED). The main results were confirmed after the exclusion of these participants.

No significant differences emerged between the participants in the excluded and included conditions regarding the main demographic and psychological characteristics considered at baseline; for more information, see Supplementary Materials.

Table 1. Demographic and clinical characteristics of the participants.

	AN $n = 42$	BN $n = 40$	BED $n = 40$	HW $n = 50$	F	p ηp^2	Post Hoc
Age, years	24.62 (8.26)	23.60 (6.86)	30.28 (9.67)	24.16 (3.41)	7.454	**<0.001** 0.117	AN < BED ($p = 0.003$) BN < BED ($p < 0.001$) HW < BED ($p = 0.001$)
BMI, kg/m^2	16.68 (0.87)	20.05 (2.06)	33.13 (10.05)	21.45 (2.96)	75.419	**<0.001** 0.574	AN < BN ($p = 0.007$) AN < BED ($p < 0.001$) AN < HW ($p < 0.001$) BN < BED ($p < 0.001$) HW < BED ($p < 0.001$)
Menarche, years	12.71 (1.29)	11.98 (1.19)	11.79 (1.27)	12.38 (1.58)	3.887	**0.010** 0.065	BED < AN ($p = 0.014$)
Education, years	13.98 (2.79)	15.21 (2.95)	14.83 (3.09)	16.06 (2.74)	4.106	**0.008** 0.069	AN < HW ($p = 0.004$)
Illness duration, months	5.83 (5.99)	5.76 (6.72)	12.14 (9.53)	-	22.638	**<0.001** 0.307	AN < BED ($p < 0.001$) BN < BED ($p < 0.001$)
PHQ-9	10.02 (4.57)	12.70 (5.75)	10.65 (4.29)	6.14 (3.94)	15.975	**<0.001** 0.222	HW < AN ($p = 0.001$) HW < BN ($p < 0.001$) HW < BED ($p < 0.001$)
EDE-Q Global	13.65 (8.23)	20.21 (8.59)	20.01 (5.36)	7.36 (6.44)	32.269	**<0.001** 0.366	HW < AN ($p < 0.001$) AN < BN ($p < 0.001$) AN < BED ($p = 0.001$) HW < BN ($p < 0.001$) HW < BED ($p < 0.001$)

Table reports means and (standard deviation), with ANOVAs' F and post hoc analyses. BMI: body mass index; AN: anorexia nervosa; BN: bulimia nervosa; BED: binge eating disorder; HW: healthy women; PHQ-9: patient health questionnaire; EDE-Q: eating disorder examination questionnaire; post hoc analysis with Bonferroni correction. Significant p values after Bonferroni correction are reported in bold.

3.1. Exclusion Condition

Means, standard deviations, and the results of the post hoc tests are reported in Table 2, while the following paragraph reports the results of the between group comparisons.

Participants with AN showed lower scores for the NTS subscales than HW, BED, and BN participants (see Table 2). The evaluation of the positive subscale of the PANAS revealed the presence of significant main effects of time ($F(1, 79) = 26.810$, $p < 0.001$) and diagnosis ($F(3, 79) = 5.169$, $p = 0.003$), as well as a significant time by diagnosis interaction ($F(3, 79) = 6.225$, $p = 0.001$). Post hoc analyses showed that HW reported higher scores than all of the ED groups and was the only group that displayed a significant decrease in positive emotions following the Cyberball task ($t(24) = 7.582$, $p < 0.001$; see Supplementary Materials for the pre and post Cyberball comparisons). A significant time by diagnosis interaction was also observed for the negative subscale of the PANAS ($F(3, 79) = 11.328$, $p < 0.001$) with HW displaying an increase in negative emotions ($t(24) = -5.630$, $p < 0.001$) and BED patients displaying a significant decrease in negative emotions ($t(19) = 2.977$, $p = 0.008$) following the exclusion condition. Regarding the SRS scale, a significant main effect of diagnosis by time emerged ($F(3, 82) = 4.467$, $p = 0.006$), with BN patients displaying the highest level of stress and BED patients the lowest. No differences due to exposure to the Cyberball task, however, were observed in any of the groups. In the TBS scale, no significant effect of diagnosis by time emerged in the change of the thoughts comparing before and after the Cyberball task ($F(3, 82) = 5.223$, $p = 0.025$). Lastly, in the TRS scale the analyses revealed the presence of significant main effects of time ($F(1, 82) = 7.167$, $p = 0.009$), and diagnosis ($F(3, 82) = 4.193$, $p = 0.008$), along with a significant time by diagnosis interaction ($F(3, 82) = 3.619$, $p = 0.013$), with both HW and BED patients displaying a significant reduction in thoughts about restriction (BED: $t(19) = 2.651$, $p = 0.016$; HW: $t(24) = 2.619$, $p = 0.015$). Post hoc tests revealed that both BN and AN patients reported significantly higher thoughts about restricting eating than HW. See Figures 2 and 3 for a graphical representation of the data.

Table 2. Need threat scale (NTS) results and comparisons.

	Exclusion Condition						
	AN *n* = 22	**BN** *n* = 19	**BED** *n* = 20	**HW** *n* = 25	**F**	*p* ηp^2	**Post Hoc**
% passes	16.70 (8.30)	15.30 (8.30)	17.10 (8.90)	21.20 (7.80)	2.181	0.096 0.074	
Belonging	9.95 (3.61)	12.53 (4.51)	13.30 (2.03)	12.48 (3.11)	3.934	**0.011** 0.126	AN < BED (*p* = **0.012**)
Self-esteem	10.50 (3.79)	11.95 (5.17)	14.20 (2.14)	14.00 (3.33)	4.922	**0.003** 0.153	AN < BED (*p* = **0.011**) AN < HW (*p* = **0.011**)
Meaningful existence	10.00 (4.27)	13.42 (4.68)	13.00 (2.43)	14.32 (3.20)	5.728	**0.001** 0.173	AN < BN (*p* = 0.026) AN < HW (*p* = **0.001**)
Control	7.68 (1.81)	8.84 (3.53)	8.85 (2.39)	9.52 (2.43)	2.024	0.117 0.069	
Manipulation check	7.41 (2.13)	6.37 (2.29)	6.90 (2.38)	6.52 (2.31)	0.890	0.450 0.032	
	Overinclusion Condition						
	AN *n* = 20	**BN** *n* = 21	**BED** *n* = 20	**HW** *n* = 25	**F**	*p* ηp^2	**Post Hoc**
% passes	67.50 (15.80)	53.90 (13.50)	72.00 (5.20)	55.80 (12.20)	10.828	**<0.001** 0.284	BN < AN (*p* = **0.004**) HW < AN (*p* = 0.013) BN < BED (*p* < **0.001**) HW< BED (*p* < **0.001**)
Belonging	20.25 (3.31)	17.81 (3.12)	21.95 (2.74)	21.16 (3.05)	7.221	**<0.001** 0.209	BN < BED (*p* < **0.001**) BN < HW (*p* = **0.002**)
Self-esteem	16.70 (4.85)	12.71 (3.86)	19.05 (2.52)	17.60 (3.49)	10.909	**<0.001** 0.285	BN < AN (*p* = **0.006**) BN < BED (*p* < **0.001**) BN < HW (*p* < **0.001**)
Meaningful existence	18.80 (4.25)	17.05 (2.04)	20.90 (2.10)	20.32 (2.75)	7.462	**<0.001** 0.214	BN < BED (*p* < **0.001**) BN < HW (*p* = **0.002**)
Control	15.95 (5.96)	13.05 (3.37)	13.45 (4.12)	16.32 (3.08)	3.483	0.019 0.113	BN < HW (*p* = 0.050)
Manipulation check	2.60 (1.31)	2.29 (0.72)	2.00 (0.00)	2.12 (0.60)	2.191	0.095 0.074	

AN: anorexia nervosa; BN: bulimia nervosa; BED: binge eating disorder; HW: healthy women; post hoc analysis with Bonferroni correction. Significant *p* values after Bonferroni correction are report-ed in bold.

3.2. Overinclusion Condition

Means, standard deviations, and the results of the post hoc tests are reported in Table 2, while the following paragraph reports the results of the between group comparisons.

The overinclusion block demonstrated the following effects: BN patients showed lower scores regarding fundamental psychological needs than both BED patients and HW, and patients with AN and BED reported a higher rate of estimated passes than patients with BN and HW (see Table 2). The analyses conducted on the positive subscale of the PANAS revealed a significant main effect of diagnosis ($F(3, 80) = 14.563$, $p < 0.001$) and time by diagnosis interaction ($F(3, 80) = 8.860$, $p < 0.001$). Post hoc tests showed that patients with BN displayed the lowest score, followed by patients with AN, and finally by patients with BED and HW, who had the same score. With regards to the interaction, while HW displayed a significant increase in positive emotions following the overinclusion condition ($t(24) = -3.565$, $p = 0.002$), patients with AN displayed a decrease ($t(19) = 2.629$, $p = 0.015$; not significant for this study). In the negative subscale of the PANAS, the analyses revealed a significant main effect of time ($F(1, 80) = 8.451$, $p = 0.005$), and diagnosis ($F(3, 80) = 9.460$, $p < 0.001$) but only a tendency to significance for time by diagnosis interaction, considering the corrected p value ($F(3, 80) = 3.580$, $p = 0.017$). The analyses conducted on the SRS scale revealed a main effect of time ($F(1, 82) = 6.554$, $p = 0.012$), a main effect of diagnosis ($F(3, 82) = 8.500$, $p < 0.001$), and a significant time by diagnosis interaction ($F(3, 82) = 3.423$,

$p = 0.021$). Post hoc testing showed that patients with BN and AN reported a higher stress level than BED patients, and only patients with BN displayed a significant reduction in the reported stress level ($t(20) = 3.782$, $p = 0.001$). Regarding the TBS scale, a significant main effect of time ($F(1, 82) = 28.654$, $p < 0.001$) and a significant time by diagnosis interaction ($F(3, 82) = 14.780$, $p < 0.001$) were found with patients with BED displaying a significant decrease in thoughts about binging ($t(19) = 4.819$, $p < 0.001$). Lastly, the analyses conducted on TRS showed only a main effect of diagnosis ($F(3, 82) = 14.307$, $p < 0.001$), with BN patients displaying higher scores on this scale. No differences due to exposure to the Cyberball task, however, were observed in any of the groups. See Figures 2 and 3 for a graphical representation.

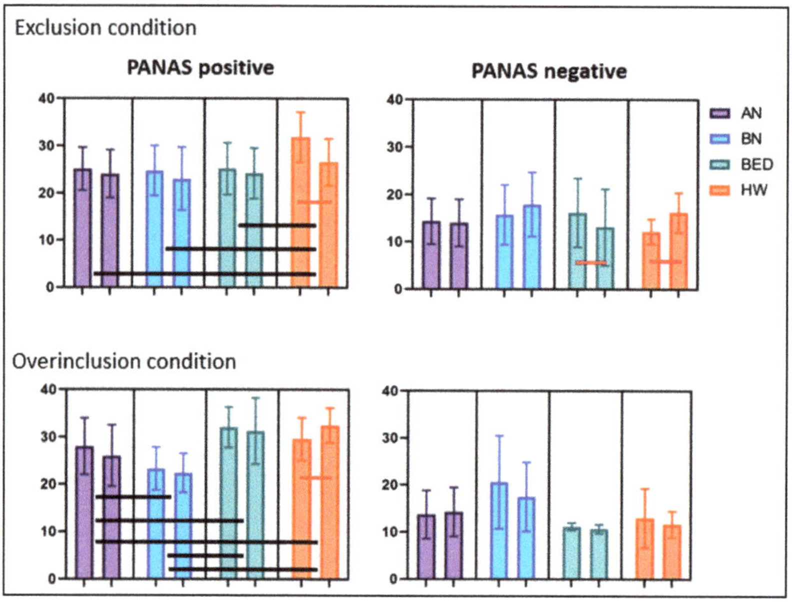

Figure 2. Graphical representation of the PANAS scores, with means and standard deviations. For all the pairs of bars, the one on the left is pre-Cyberball, and the one on the right is post-Cyberball. The bars show the presence of a significant difference between pre and post comparisons (red bars) or between diagnoses (black bars).

3.3. Conditions Comparison

Looking at the effects between exclusion and overinclusion conditions, we found significant differences for all the NTS subscales and the percentage of the ball passes, with participants who reported higher scores after overinclusion condition ($p < 0.001$ for all the subscales—belonging, self-esteem, meaningful existence, control, and manipulation check—and $p = 0.003$ for the percentage of the ball passes). Looking at the interaction between conditions and groups, we found significant differences for belonging ($F(3, 164) = 4.308$, $p = 0.006$, $\eta p^2 = 0.860$), self-esteem ($F(3, 164) = 3.896$, $p = 0.010$, $\eta p^2 = 0.820$), meaningful existence ($F(3, 164) = 4.800$, $p = 0.003$, $\eta p^2 = 0.897$), the percentage of the ball passes ($F(3, 164) = 9.252$, $p < 0.001$, $\eta p^2 = 0.996$), and a tendency to being significant for the control ($F(3, 164) = 3.109$, $p = 0.028$, $\eta p^2 = 0.717$), but no differences in regard to the manipulation check ($F(3, 164) = 0.502$, $p = 0.681$, $\eta p^2 = 0.151$). Looking at pairwise comparisons, significant differences emerged between BN and BED ($p = 0.001$), and BED and HW ($p = 0.037$), for the percentage of the ball passes. For belonging, significant differences emerged between AN and BED ($p = 0.001$), AN and HW ($p = 0.022$), and BN and BED ($p = 0.009$). For self-esteem, significant differences emerged between AN and BED ($p = 0.001$), AN and HW ($p = 0.016$),

BN and BED ($p < 0.001$), and BN and HW ($p < 0.001$). Finally, for meaningful existence, significant differences emerged between AN and BED ($p = 0.001$), AN and HW ($p < 0.001$), and BN and HW ($p = 0.027$).

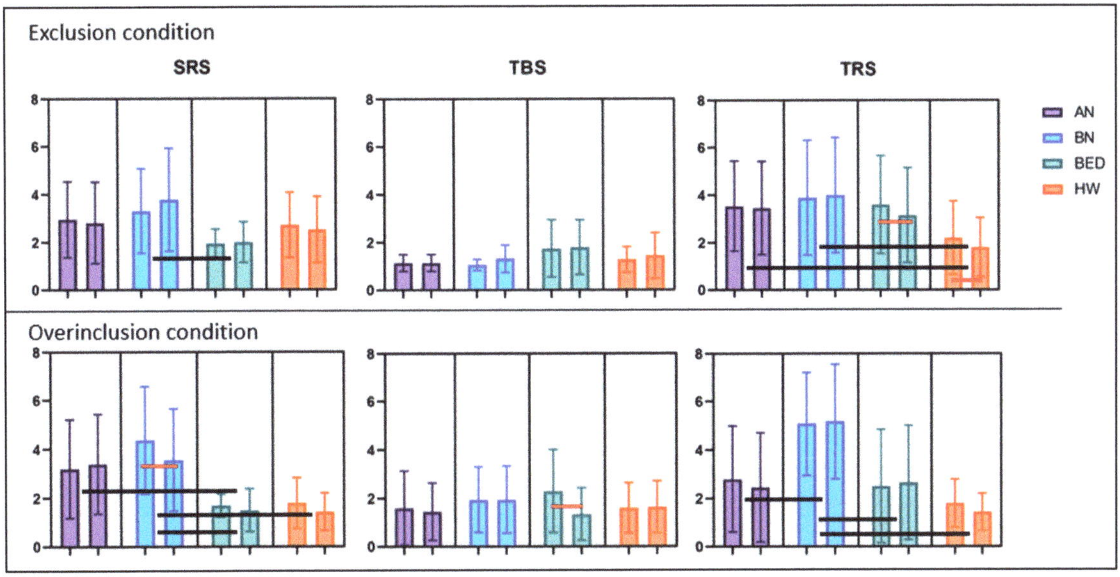

Figure 3. Graphical representation of the SRS, TBS, and TRS in the Cyberball conditions, with means and standard deviations. For all the pairs of bars, the one on the left is pre-Cyberball, and the one on the right is post-Cyberball. The bars show the presence of a significant difference between pre and post (red bars) comparisons or between diagnoses (black bars).

Looking at the interactions of time, conditions, and diagnoses, we found several significant main effects. The PANAS positive subscale showed a significant interaction ($F(3, 159) = 14.686$, $p < 0.001$), with significant comparisons between diagnoses at the post hoc analysis: AN $\neq$ HW with $p < 0.001$, BN $\neq$ BED with $p < 0.001$, and BN $\neq$ HW with $p < 0.001$. The same was for the PANAS negative subscale, which showed a significant interaction ($F(3, 159) = 11.331$, $p < 0.001$) with specific differences between diagnoses: AN $\neq$ BN with $p = 0.009$, BN $\neq$ BED with $p < 0.001$, and BN $\neq$ HW with $p < 0.001$. Looking at the SRS scale, we found a significant interaction ($F(3, 164) = 3.444$, $p = 0.013$) with specific differences between diagnoses: AN $\neq$ BED with $p = 0.001$, BN $\neq$ BED with $p < 0.001$, and BN $\neq$ HW with $p = 0.002$. The TBS scales also showed significant interaction ($F(3, 164) = 10.740$, $p < 0.001$) with only one significant difference at the post hoc analysis: AN $\neq$ BED with $p = 0.036$. Differently, no significant interaction was found for the TRS scale ($F(3, 164) = 1.411$, $p = 0.241$).

3.4. Correlations

In the excluded condition, correlation analyses found that while both HW and AN patients showed no significant results among the included psychological variables; the BN and BED groups showed negative correlations between fundamental needs (NTS scores) evaluated after the Cyberball task and the modification of their emotional states. Moreover, BN patients showed a negative correlation between the meaningful existence subscale and the change in their stress level ($r = -0.696$), while BED patients showed a negative correlation between the belonging NTS subscale and changes in their thoughts about restriction level ($r = -0.728$). In the overinclusion condition, BN patients showed negative correlations between changes in all of the thoughts scores and part of the NTS scores (meaningful existence and DSRS, $r = -0.606$; belonging and DTBS, $r = -0.690$; control and DTBS, $r = -0.564$; self-esteem and DTRS, $r = -0.610$), while BED patients showed positive correlations (belonging and DSRS $r = 0.612$; self-esteem and DSRS, $r = 0.757$; meaningful existence and DSRS, $r = 0.734$; control and DSRS, $r = 0.717$; control and DTBS, $r = 0.604$; self-esteem and DTRS, $r = 0.738$). No correlations emerged for the HW and AN groups. See Supplementary Materials for details. Regarding age at menarche, no significant correlation emerged in either condition or any of the groups.

4. Discussion

The current study examined changes in emotions, stress, and eating disorder attitudes after a simulated social overinclusion or exclusion scenario via a ball-tossing game. This is the first study that examined the effect of social exclusion and overinclusion across the ED spectrum and aimed to extend the previous findings in patients with AN, looking for similarities or differences between the entire spectrum of patients with ED.

Evaluation of the perception of ball passes showed that all the subgroups could discriminate between exclusion and overinclusion in the ball-toss game; our results confirmed preliminary evidence about the ability of AN patients to determine the nature of a social scenario [1,41] and added new data for patients with BN and BED. For participants' perception of their overinclusion or exclusion in the game, more significant differences emerged during the overincluding paradigm, confirming that social comparison is a vulnerability aspect that should be carefully considered when treating ED patients [52].

Some peculiarities concerning the specific ED diagnosis also emerged. Patients with AN demonstrated psychological difficulty in being excluded by peers with lower scores at the evaluation after the Cyberball task [41], while BN patients showed that overinclusion by peers could be more complex to manage.

Indeed, patients with BN reported that interpersonal exposure induced by the over including condition results in lower scores for self-esteem and sense of belonging than other participants. Both of these elements agree with the interpersonal model for ED [15], which shows that AN patients have more fear of negative evaluations, while BN patients have higher rates of negative interactions with others. Our results align with the impaired social effects of interpersonal interaction that are linked, in BN patients, more to the reduction of negative emotions than to an increase in positive ones (see above). Our results highlighted different risk factors in the interpersonal domain for each clinical group, demonstrating specific psychological needs that should also be considered in the therapeutic process [53], especially looking at personalized approaches [54].

In line with our previous study, the data indicate difficulties in emotional reactivity in AN patients after being ostracized [41], reporting a higher negative impact on self-esteem, sense of belonging, and psychological well-being. The new data, however, show that this aspect is a key feature characterizing the entire ED spectrum, with no differences among ED diagnoses (AN, BN, BED). The difficulties in understanding their emotional states may suggest a possible cognitive factor that has a role in developing and maintaining dysfunctional behaviors across the entire spectrum [52,55]. Indeed, looking at the affective response to the Cyberball paradigm, no significant differences emerged between clinical groups in the excluded condition. Moreover, significant changes in the emotional state were reported only by healthy individuals. This aspect might underline an impairment in ED patients' ability to understand their emotional states linked to interpersonal interactions, in line with dysfunctional emotional reactiveness and interpersonal models of EDs [56]. Patients might present a reduced emotional reactivity to this kind of social interaction or task, corroborating the previous data about the possible presence of impaired emotional reactivity in patients with an ED [57]. Indeed, different studies have pointed out the possible role of emotion regulation in EDs, showing the relevance of targeting emotional skills in treatments [58,59]. However, another possible explanation is the presence of high alexithymia, a severe trait present in patients with EDs, or dissociation during the task due to a non-acceptance of emotional changes [60,61].

Interestingly, patients with BED reduced their negative emotional state after being excluded, which was linked to the high level of belonging reported on the NTS scale. The same results have been found in post-bariatric patients [62], where the same correlations were found, showing a similarity in interpersonal skills difficulties [63] that might be explained by a shared impaired capacity to understand and code their own emotions [21]. Indeed, although the link between precipitating negative affect and binging is already well documented, little is known about the cognitive or behavioral chain; moreover, recent evidence has demonstrated the importance of intra-individual variability in the precipitating affective processes and the control of food-related impulsivity [64]. Another possible explanation might be the presence of an emotional detachment response to acute social exclusion that has been proposed as an "invisible shield", with an automatic emotion process in which positive emotions become highly accessible [65]. This process might be true for people with high social negative evaluations and exclusion [23,66], such as individuals with BED [67]. For these reasons, future studies might focus on emotional processing in individuals with BED, and could evaluate the implementation of emotional recognition and interpretation in a psychological approach that has already shown interesting effects in the reduction of food impulsivity [68,69], looking at personalized and improved treatment approaches [70].

In the overincluding paradigm, patients with BN showed a significant decrease in negative emotions, with a possible positive effect of social interactions on negative emotions but not on positive emotions that may be linked to an impaired reward from social interactions [71]. Interestingly, even if overinclusion reduced stress levels in BN patients, no differences emerged regarding eating thoughts, corroborating the evidence that patients with BN could present a dissociation between emotional states and psychopathological behavioral responses [28].

Several studies in the literature have pointed out the role of the age of menarche in emotional regulation and social cognition [72–74]. Our data do not support this aspect, but the heterogeneity of the sample and the specificity of the relationship between EDs and menarche might be considered for future evaluations. Indeed, bodily changes after menarche are possible stressors for developing an ED, impacting the physiological relationships between the hormonal cycle and the development of cognitive functions such as emotional regulation due to dysfunctional eating behaviors [75–77].

No relationships emerged between psychological need, psychopathology, and eating thoughts in the control group, which differed from previous literature and may be linked to the different tasks applied [35]. Overall, the HW group reported the largest modulation of emotional states in this experiment, corroborating the idea that individuals with EDs apply maladaptive strategies in interpersonal emotion regulation [19]. These data are in line with the possible role of rumination and thought suppression as cognitive dysfunctional emotion regulation strategies in individuals with ED [78]. Rumination—an obsessive and passive focus on the meaning, causes, and consequences of emotions—and suppression thereof—disengagement from unwanted thoughts—are maladaptive strategies to disengage from the context. They have been related to dysfunctional eating behaviors in patients, with a paradoxical increase in negative emotions and cognitions [78] that might also be a difficult trait in patients [79].

Although the relevance of interpersonal relationships has been noted in several treatment approaches, from interpersonal therapy to family-based therapy and the Maudsley Model (MANTRA), more studies are needed [19]. Emotional regulation has been shown to have a relevant role in the treatment pathway, possibly impacting the therapeutic alliance and recovery. Our data showed a dynamic response modulation in patients with AN, suppression of emotional changes and increased eating behaviors in BN patients via social manipulation, and difficulties in emotional self-reading in BED patients linked to social cues. This indicates specific differences regarding social cognition in the ED spectrum that, as suggested by a recent literature review, require more study [1]. These aspects might also be important in different scenarios and have been pointed out as crucial, for example, during the COVID-19 pandemic [80]. Indeed, the disruption of social connections and routines has been related to more severe psychopathology, and impaired coping strategies in social situations might explain different responses to the pandemic [81–83], calling for new studies in the field.

This is the first study to apply the Cyberball task to the whole spectrum of EDs, and the main strength is the experimental psychopathological approach applied, but upon examining our results, several limitations should be considered. First, the self-reporting nature of the data could affect the self-report results, and future studies should include different ways to record participants' responses (e.g., skin conductance, heartbeat). Second, in seeking increasing homogeneity, we included only female participants, but this choice could limit the generalizability of the results. Third, this is a laboratory-based experiment. Even if none of the participants were familiar with the Cyberball task before the session or reported doubts about the real presence of the other participants, more studies are needed to confirm our results and to compare a laboratory task to real-world situations. Fourth, our results need to be replicated and confirmed with larger populations and from different experimental studies. Finally, in evaluating the psychological effects of the Cyberball task, we included participants of different ages, but we may have lost some specific differences between adolescents and adults that could be investigated in future studies.

5. Conclusions

Despite some limitations, this is the first study to evaluate the effect of social exclusion and overinclusion across the ED spectrum. Our results highlight the relevant role of the interpersonal domain in EDs, showing possible trigger mechanisms in either overinclusion or exclusion that could directly affect eating behaviors. Indeed, participants from the general population shown the largest emotional modulation considering both conditions, while patients with AN showed a different vulnerability toward being ostracized and overincluded than those with BN or BED, suggesting that more studies are needed in order to comprehend patients' ability to understand social, emotional, and interpersonal scenarios [1,20]. Interpersonal emotion regulation and its relationship with eating behaviors in patients with EDs should be evaluated with different laboratory tasks, examining possible differences due to the duration of the disorder or age of the participants (adolescence vs. adulthood). Future studies should include different psychological constructs such as social anxiety and fear of positive/negative evaluations, seeking improved differentiations that could help clinicians create personalized interventions. Indeed, the research should look for specific training that could improve the coping strategies and treatment outcomes proposed in the literature, from the cognitive approach to emotional cues to approaches more linked to the third generation of cognitive therapies [84,85].

Supplementary Materials: The following supporting information can be downloaded at: https://www.mdpi.com/article/10.3390/nu15041021/s1. Table S1: Comparisons between participants in the excluded and overincluded conditions at baseline; Table S2: Correlation analyses exclusion condition; Table S3: Correlation analyses overinclusion condition; Table S4: Evaluation of change due to the Cyberball task.

Author Contributions: Conceptualization, P.M., E.T. and A.F.; methodology, P.M.; software, P.M.; validation, P.M., V.M. and E.C.; formal analysis, P.M. and V.M.; investigation, P.M.; resources, A.F.; data curation, P.M.; writing—original draft preparation, P.M., V.M. and E.C.; writing—review and editing, V.C., E.T. and A.F.; visualization, P.M.; supervision, E.T. and A.F.; project administration, A.F.; funding acquisition, A.F. All authors have read and agreed to the published version of the manuscript.

Funding: This work was supported by the "Department of excellence 2018–2022" initiative of the Italian Ministry of Education (MIUR) awarded to the Department of Neuroscience—University of Padua.

Institutional Review Board Statement: All procedures performed in studies involving human participants were in accordance with the ethical standards of the institutional and/or national research committee and the 1964 Declaration of Helsinki and its later amendments or comparable ethical standards. The study was approved by the local Ethics Committee for Clinical Practice (protocol number: 1469/2018) on 13 March 2018.

Informed Consent Statement: Informed consent was obtained from all subjects involved in the study.

Data Availability Statement: The datasets used and analyzed during the current study are available from the corresponding author upon reasonable request.

Conflicts of Interest: The authors declare no conflict of interest.

References

1. Mason, T.B.; Lesser, E.L.; Dolgon-Krutolow, A.R.; Wonderlich, S.A.; Smith, K.E. An Updated Transdiagnostic Review of Social Cognition and Eating Disorder Psychopathology. *J. Psychiatr. Res.* **2021**, *143*, 602–627. [CrossRef]
2. Meneguzzo, P.; Todisco, P.; Calonaci, S.; Mancini, C.; Dal Brun, D.; Collantoni, E.; Donini, L.M.; Tenconi, E.; Favaro, A. Health-Related Quality of Life Assessment in Eating Disorders: Adjustment and Validation of a Specific Scale with the Inclusion of an Interpersonal Domain. *Eat. Weight. Disord.* **2020**, *26*, 2251–2262. [CrossRef] [PubMed]
3. Caglar-Nazali, H.P.; Corfield, F.; Cardi, V.; Ambwani, S.; Leppanen, J.; Olabintan, O.; Deriziotis, S.; Hadjimichalis, A.; Scognamiglio, P.; Eshkevari, E.; et al. A Systematic Review and Meta-Analysis of "Systems for Social Processes" in Eating Disorders. *Neurosci. Biobehav. Rev.* **2014**, *42*, 55–92. [CrossRef] [PubMed]
4. Corsi, E.; Cardi, V.; Sowden, S.; Coll, M.P.; Cascino, G.; Ricca, V.; Treasure, J.; Bird, G.; Monteleone, A.M. Socio-Cognitive Processing in People with Eating Disorders: Computerized Tests of Mentalizing, Empathy and Imitation Skills. *Int. J. Eat. Disord.* **2021**, *54*, 1509–1518. [CrossRef] [PubMed]
5. Marzola, E.; Porliod, A.; Panero, M.; De-Bacco, C.; Abbate-Daga, G. Affective Temperaments and Eating Psychopathology in Anorexia Nervosa: Which Role for Anxious and Depressive Traits? *J. Affect. Disord.* **2020**, *266*, 374–380. [CrossRef] [PubMed]
6. Monteleone, A.M.; Treasure, J.; Kan, C.; Cardi, V. Reactivity to Interpersonal Stress in Patients with Eating Disorders: A Systematic Review and Meta-Analysis of Studies Using an Experimental Paradigm. *Neurosci. Biobehav. Rev.* **2018**, *87*, 133–150. [CrossRef] [PubMed]
7. McFillin, R.K.; Cahn, S.C.; Burks, V.S.; Levine, M.P.; Loney, S.L.; Levine, R.L. Social Information-Processing and Coping in Adolescent Females Diagnosed With an Eating Disorder: Toward a Greater Understanding of Control. *Eat. Disord.* **2012**, *20*, 42–59. [CrossRef] [PubMed]
8. Bronstein, M.V.; Everaert, J.; Kummerfeld, E.; Haynos, A.F.; Vinogradov, S. Biased and Inflexible Interpretations of Ambiguous Social Situations: Associations with Eating Disorder Symptoms and Socioemotional Functioning. *Int. J. Eat. Disord.* **2022**, *55*, 518–529. [CrossRef]
9. Levinson, C.A.; Rodebaugh, T.L. Social Anxiety and Eating Disorder Comorbidity: The Role of Negative Social Evaluation Fears. *Eat. Behav.* **2012**, *13*, 27–35. [CrossRef]
10. Schaumberg, K.; Reilly, E.E.; Gorrell, S.; Levinson, C.A.; Farrell, N.R.; Brown, T.A.; Smith, K.M.; Schaefer, L.M.; Essayli, J.H.; Haynos, A.F.; et al. Conceptualizing Eating Disorder Psychopathology Using an Anxiety Disorders Framework: Evidence and Implications for Exposure-Based Clinical Research. *Clin. Psychol. Rev.* **2021**, *83*, 101952. [CrossRef]
11. Menatti, A.R.; DeBoer, L.B.H.; Weeks, J.W.; Heimberg, R.G. Social Anxiety and Associations with Eating Psychopathology: Mediating Effects of Fears of Evaluation. *Body Image* **2015**, *14*, 20–28. [CrossRef] [PubMed]
12. Cook, S.I.; Moore, S.; Bryant, C.; Phillips, L.J. The Role of Fear of Positive Evaluation in Social Anxiety: A Systematic Review and Meta-Analysis. *Clin. Psychol. Sci. Pract.* **2022**, *29*, 352. [CrossRef]
13. Levinson, C.A.; Sala, M.; Murray, S.; Ma, J.; Rodebaugh, T.L.; Lenze, E.J. Diagnostic, Clinical, and Personality Correlates of Food Anxiety during a Food Exposure in Patients Diagnosed with an Eating Disorder. *Eat. Weight. Disord.* **2019**, *24*, 1079–1088. [CrossRef] [PubMed]
14. Robinson, K.; Wade, T.D. Perfectionism Interventions Targeting Disordered Eating: A Systematic Review and Meta-Analysis. *Int. J. Eat. Disord.* **2021**, *54*, 473–487. [CrossRef]
15. Arcelus, J.; Haslam, M.; Farrow, C.; Meyer, C. The Role of Interpersonal Functioning in the Maintenance of Eating Psychopathology: A Systematic Review and Testable Model. *Clin. Psychol. Rev.* **2013**, *33*, 156–167. [CrossRef]
16. Brockmeyer, T.; Walther, S.; Ingenerf, K.; Wild, B.; Hartmann, M.; Weisbrod, M.; Weber, M.A.; Eckhardt-Henn, A.; Herzog, W.; Friederich, H.C. Brain Effects of Computer-Assisted Cognitive Remediation Therapy in Anorexia Nervosa: A Pilot FMRI Study. *Psychiatry Res. Neuroimaging* **2016**, *249*, 52–56. [CrossRef]
17. Carter, J.C.; Mercer-Lynn, K.B.; Norwood, S.J.; Bewell-Weiss, C.V.; Crosby, R.D.; Woodside, D.B.; Olmsted, M.P. A Prospective Study of Predictors of Relapse in Anorexia Nervosa: Implications for Relapse Prevention. *Psychiatry Res.* **2012**, *200*, 518–523. [CrossRef]
18. Courty, A.; Godart, N.; Lalanne, C.; Berthoz, S. Alexithymia, a Compounding Factor for Eating and Social Avoidance Symptoms in Anorexia Nervosa. *Compr. Psychiatry* **2015**, *56*, 217–228. [CrossRef]

19. Christensen, K.A.; Haynos, A.F. A Theoretical Review of Interpersonal Emotion Regulation in Eating Disorders: Enhancing Knowledge by Bridging Interpersonal and Affective Dysfunction. *J. Eat. Disord.* **2020**, *8*, 21. [CrossRef]
20. Bora, E.; Köse, S. Meta-Analysis of Theory of Mind in Anorexia Nervosa and Bulimia Nervosa: A Specific İmpairment of Cognitive Perspective Taking in Anorexia Nervosa? *Int. J. Eat. Disord.* **2016**, *49*, 739–740. [CrossRef]
21. Aloi, M.; Rania, M.; Caroleo, M.; De Fazio, P.; Segura-García, C. Social Cognition and Emotional Functioning in Patients with Binge Eating Disorder. *Eur. Eat. Disord. Rev.* **2017**, *25*, 172–178. [CrossRef] [PubMed]
22. Westwood, H.; Kerr-Gaffney, J.; Stahl, D.; Tchanturia, K. Alexithymia in Eating Disorders: Systematic Review and Meta-Analyses of Studies Using the Toronto Alexithymia Scale. *J. Psychosom. Res.* **2017**, *99*, 66–81. [CrossRef] [PubMed]
23. Vartanian, L.R.; Porter, A.M. Weight Stigma and Eating Behavior: A Review of the Literature. *Appetite* **2016**, *102*, 3–14. [CrossRef] [PubMed]
24. Cardi, V.; Turton, R.; Schifano, S.; Leppanen, J.; Hirsch, C.R.; Treasure, J. Biased Interpretation of Ambiguous Social Scenarios in Anorexia Nervosa. *Eur. Eat. Disord. Rev.* **2017**, *25*, 60–64. [CrossRef]
25. Reinhard, M.A.; Dewald-Kaufmann, J.; Wüstenberg, T.; Musil, R.; Barton, B.B.; Jobst, A.; Padberg, F. The Vicious Circle of Social Exclusion and Psychopathology: A Systematic Review of Experimental Ostracism Research in Psychiatric Disorders. *Eur. Arch. Psychiatry Clin. Neurosci.* **2020**, *270*, 521–532. [CrossRef]
26. Baumeister, R.F.; Brewer, L.E.; Tice, D.M.; Twenge, J.M. Thwarting the Need to Belong: Understanding the Interpersonal and Inner Effects of Social Exclusion. *Soc. Pers. Psychol. Compass* **2007**, *1*, 506–520. [CrossRef]
27. Salvy, S.J.; de la Haye, K.; Bowker, J.C.; Hermans, R.C.J. Influence of Peers and Friends on Children's and Adolescents' Eating and Activity Behaviors. *Physiol. Behav.* **2012**, *106*, 369–378. [CrossRef]
28. Tuschen-Caffier, B.; Vögele, C. Psychological and Physiological Reactivity to Stress: An Experimental Study on Bulimic Patients, Restrained Eaters and Controls. *Psychother. Psychosom.* **1999**, *68*, 333–340. [CrossRef]
29. Dalgleish, T.; Walsh, N.D.; Mobbs, D.; Schweizer, S.; Van Harmelen, A.L.; Dunn, B.; Dunn, V.; Goodyer, I.; Stretton, J. Social Pain and Social Gain in the Adolescent Brain: A Common Neural Circuitry Underlying Both Positive and Negative Social Evaluation. *Sci. Rep.* **2017**, *7*, 42010. [CrossRef]
30. Fredrick, J.W.; Luebbe, A.M. Fear of Positive Evaluation and Social Anxiety: A Systematic Review of Trait-Based Findings. *J. Affect. Disord.* **2020**, *265*, 157–168. [CrossRef]
31. Williams, K.D.K.D.; Jarvis, B. Cyberball: A Program for Use in Research on Interpersonal Ostracism and Acceptance. *J. Behav. Res. Methods* **2006**, *38*, 174–180. [CrossRef] [PubMed]
32. Heeren, A.; Dricot, L.; Billieux, J.; Philippot, P.; Grynberg, D.; De Timary, P.; Maurage, P. Correlates of Social Exclusion in Social Anxiety Disorder: An FMRI Study. *Sci. Rep.* **2017**, *7*, 260. [CrossRef] [PubMed]
33. Seidl, E.; Padberg, F.; Bauriedl-Schmidt, C.; Albert, A.; Daltrozzo, T.; Hall, J.; Renneberg, B.; Seidl, O.; Jobst, A. Response to Ostracism in Patients with Chronic Depression, Episodic Depression and Borderline Personality Disorder a Study Using Cyberball. *J. Affect. Disord.* **2020**, *260*, 254–262. [CrossRef] [PubMed]
34. Sebastian, C.; Blakemore, S.J.; Charman, T. Reactions to Ostracism in Adolescents with Autism Spectrum Conditions. *J. Autism. Dev. Disord.* **2009**, *39*, 1122–1130. [CrossRef] [PubMed]
35. Schell, S.E.; Banica, I.; Weinberg, A.; Racine, S.E. Hunger Games: Associations between Core Eating Disorder Symptoms and Responses to Rejection by Peers during Competition. *Int. J. Eat. Disord.* **2021**, *54*, 802–811. [CrossRef]
36. Senese, V.P.; Pezzella, M.; Pasquariello, L.; Ali, S.; Rohner, R.P. Effects of Social Exclusion and Maternal Rejection on Children's High-Caloric Food Consumption. *Appetite* **2020**, *145*, 104494. [CrossRef]
37. Salvy, S.J.; Bowker, J.C.; Nitecki, L.A.; Kluczynski, M.A.; Germeroth, L.J.; Roemmich, J.N. Impact of Simulated Ostracism on Overweight and Normal-Weight Youths' Motivation to Eat and Food Intake. *Appetite* **2011**, *56*, 39–45. [CrossRef]
38. Weiss, M.; Huppert, J.D. Shy, but Why? Vulnerable Narcissism and Avoidant Personality in Terms of Explicit and Implicit Interpretation Bias and Social Acceptance. *Curr. Psychol.* **2022**, *2022*, 1–13. [CrossRef]
39. Karp, J.N.; Makol, B.A.; Keeley, L.M.; Qasmieh, N.; Deros, D.E.; Weeks, J.W.; Racz, S.J.; Lipton, M.F.; Augenstein, T.M.; De Los Reyes, A. Convergent, Incremental, and Criterion-Related Validity of Multi-Informant Assessments of Adolescents' Fears of Negative and Positive Evaluation. *Clin. Psychol. Psychother.* **2018**, *25*, 217–230. [CrossRef]
40. Tanaka, H.; Ikegami, T. Fear of Negative Evaluation Moderates Effects of Social Exclusion on Selective Attention to Social Signs. *Cogn. Emot.* **2015**, *29*, 1306–1313. [CrossRef]
41. Meneguzzo, P.; Collantoni, E.; Bonello, E.; Busetto, P.; Tenconi, E.; Favaro, A. The Predictive Value of the Early Maladaptive Schemas in Social Situations in Anorexia Nervosa. *Eur. Eat. Disord. Rev.* **2020**, *28*, 318–331. [CrossRef] [PubMed]
42. American Psychiatric Association. *Diagnostic and Statistical Manual of Mental Disorders (DSM-5)*; American Psychiatric Association: Washington, DC, USA, 2013.
43. Fairburn, C.G.; Beglin, S.J. Assessment of Eating Disorder: Interview or Self-Report Questionnaire? *Int. J. Eat. Disord.* **1994**, *16*, 363–370. [CrossRef] [PubMed]
44. Kroenke, K.; Spitzer, R.L.; Williams, J.B.W.W. The PHQ-9: Validity of a Brief Depression Severity Measure. *J. Gen. Intern. Med.* **2001**, *16*, 606–613. [CrossRef] [PubMed]
45. Watson, D.; Clark, L.A.; Tellegen, A. Development and Validation of Brief Measures of Positive and Negative Affect: The PANAS Scales. *J. Pers. Soc. Psychol.* **1988**, *54*, 1063–1070. [CrossRef]

46. Gerber, J.P.; Chang, S.-H.; Reimel, H. Construct Validity of Williams' Ostracism Needs Threat Scale. *Pers. Individ. Dif.* **2017**, *115*, 50–53. [CrossRef]
47. Zadro, L.; Williams, K.D.; Richardson, R. How Low Can You Go? Ostracism by a Computer Is Sufficient to Lower Self-Reported Levels of Belonging, Control, Self-Esteem, and Meaningful Existence. *J. Exp. Soc. Psychol.* **2004**, *40*, 560–567. [CrossRef]
48. Hawes, D.J.; Zadro, L.; Fink, E.; Richardson, R.; O'Moore, K.; Griffiths, B.; Dadds, M.R.; Williams, K.D. The Effects of Peer Ostracism on Children's Cognitive Processes. *Eur. J. Dev. Psychol.* **2012**, *9*, 599–613. [CrossRef]
49. De Panfilis, C.; Riva, P.; Preti, E.; Cabrino, C.; Marchesi, C. When Social Inclusion Is Not Enough: Implicit Expectations of Extreme Inclusion in Borderline Personality Disorder. *Personal. Disord. Theory Res. Treat.* **2015**, *6*, 301. [CrossRef]
50. Weinbrecht, A.; Niedeggen, M.; Roepke, S.; Renneberg, B. Feeling Excluded No Matter What? Bias in the Processing of Social Participation in Borderline Personality Disorder. *Neuroimage Clin.* **2018**, *19*, 343–350. [CrossRef]
51. Wendler, D.; Miller, F.G. Deception in the Pursuit of Science. *Arch. Intern. Med.* **2004**, *164*, 597–600. [CrossRef]
52. Treasure, J.; Corfield, F.; Cardi, V. A Three-Phase Model of the Social Emotional Functioning in Eating Disorders. *Eur. Eat. Disord. Rev.* **2012**, *20*, 431–438. [CrossRef] [PubMed]
53. Björck, C.; Clinton, D.; Sohlberg, S.; Hällström, T.; Norring, C. Interpersonal Profiles in Eating Disorders: Ratings of SASB Self-Image. *Psychol. Psychother. Theory Res. Pract.* **2003**, *76*, 337–349. [CrossRef] [PubMed]
54. Treasure, J. Treatment Resistance in Eating Disorders: A Question of Uncertain Targets or Insufficient Personalization? *Int. Rev. Psychiatry* **2019**, *31*, 305–307. [CrossRef] [PubMed]
55. Anderson, L.K.; Claudat, K.; Cusack, A.; Brown, T.A.; Trim, J.; Rockwell, R.; Nakamura, T.; Gomez, L.; Kaye, W.H. Differences in Emotion Regulation Difficulties among Adults and Adolescents across Eating Disorder Diagnoses. *J. Clin. Psychol.* **2018**, *74*, 1867–1873. [CrossRef]
56. De Paoli, T.; Fuller-Tyszkiewicz, M.; Krug, I. Insecure Attachment and Maladaptive Schema in Disordered Eating: The Mediating Role of Rejection Sensitivity. *Clin. Psychol. Psychother.* **2017**, *24*, 1273–1284. [CrossRef]
57. De Sampaio, F.T.P.; Soneira, S.; Aulicino, A.; Harris, P.; Allegri, R.F. Emotional Reactivity to Social Stimuli in Patients with Eating Disorders. *Psychiatry Res.* **2015**, *229*, 887–894. [CrossRef]
58. Mallorquí-Bagué, N.; Vintró-Alcaraz, C.; Sánchez, I.; Riesco, N.; Agüera, Z.; Granero, R.; Jiménez-Múrcia, S.; Menchón, J.M.; Treasure, J.; Fernández-Aranda, F. Emotion Regulation as a Transdiagnostic Feature Among Eating Disorders: Cross-Sectional and Longitudinal Approach. *Eur. Eat. Disord. Rev.* **2018**, *26*, 53–61. [CrossRef]
59. Danner, U.N.; Sternheim, L.; Evers, C. The Importance of Distinguishing between the Different Eating Disorders (Sub) Types When Assessing Emotion Regulation Strategies. *Psychiatry Res.* **2014**, *215*, 727–732. [CrossRef]
60. Meneguzzo, P.; Garolla, A.; Bonello, E.; Todisco, P. Alexithymia, Dissociation, and Emotional Regulation in Eating Disorders: Evidence of Improvement through Specialized Inpatient Treatment. *Clin. Psychol. Psychother.* **2021**, *29*, 718–724. [CrossRef]
61. Longo, P.; Panero, M.; Amodeo, L.; Demarchi, M.; Abbate-Daga, G.; Marzola, E. Psychoform and Somatoform Dissociation in Anorexia Nervosa: A Systematic Review. *Clin. Psychol. Psychother.* **2021**, *28*, 295–312. [CrossRef]
62. Meneguzzo, P.; Tenconi, E.; Collantoni, E.; Longobardi, G.; Zappalà, A.; Vindigni, V.; Favaro, A.; Pavan, C. The Cyberball Task in People after Obesity Surgery: Preliminary Evaluation of Cognitive Effects of Social Inclusion and Exclusion with a Laboratory Task. *Eat. Weight. Disord.* **2021**, *27*, 1523–1533. [CrossRef] [PubMed]
63. Duchesne, M.; De Oliveira Falcone, E.M.; De Freitas, S.R.; D'Augustin, J.F.; Marinho, V.; Appolinario, J.C. Assessment of Interpersonal Skills in Obese Women with Binge Eating Disorder. *J. Health Psychol.* **2012**, *17*, 1065–1075. [CrossRef] [PubMed]
64. Smith, K.E.; Mason, T.B.; Schaefer, L.M.; Juarascio, A.; Dvorak, R.; Weinbach, N.; Crosby, R.D.; Wonderlich, S.A. Examining Intra-Individual Variability in Food-Related Inhibitory Control and Negative Affect as Predictors of Binge Eating Using Ecological Momentary Assessment. *J. Psychiatr. Res.* **2020**, *120*, 137–143. [CrossRef] [PubMed]
65. DeWall, C.N.; Twenge, J.M.; Koole, S.L.; Baumeister, R.F.; Marquez, A.; Reid, M.W. Automatic Emotion Regulation After Social Exclusion: Tuning to Positivity. *Emotion* **2011**, *11*, 623–636. [CrossRef] [PubMed]
66. Almeida, L.; Savoy, S.; Boxer, P. The Role of Weight Stigmatization in Cumulative Risk for Binge Eating. *J. Clin. Psychol.* **2011**, *67*, 278–292. [CrossRef] [PubMed]
67. Sawaoka, T.; Barnes, R.D.; Blomquist, K.K.; Masheb, R.M.; Grilo, C.M. Social Anxiety and Self-Consciousness in Binge Eating Disorder: Associations with Eating Disorder Psychopathology. *Compr. Psychiatry* **2012**, *53*, 740–745. [CrossRef]
68. Dingemans, A.; Danner, U.; Parks, M. Emotion Regulation in Binge Eating Disorder: A Review. *Nutrients* **2017**, *9*, 1274. [CrossRef]
69. Schag, K.; Leehr, E.J.; Meneguzzo, P.; Martus, P.; Zipfel, S.; Giel, K.E. Food-Related Impulsivity Assessed by Longitudinal Laboratory Tasks Is Reduced in Patients with Binge Eating Disorder in a Randomized Controlled Trial. *Sci. Rep.* **2021**, *11*, 8225. [CrossRef]
70. Clyne, C.; Latner, J.D.; Gleaves, D.H.; Blampied, N.M. Treatment of Emotional Dysregulation in Full Syndrome and Subthreshold Binge Eating Disorder. *Eat. Disord.* **2010**, *18*, 408–424. [CrossRef]
71. Barkus, E.; Badcock, J.C. A Transdiagnostic Perspective on Social Anhedonia. *Front. Psychiatry* **2019**, *10*, 216. [CrossRef]
72. Alcalá-Herrera, V.; Marván, M.L. Early Menarche, Depressive Symptoms, and Coping Strategies. *J. Adolesc.* **2014**, *37*, 905–913. [CrossRef] [PubMed]
73. Barrau-Sastre, P.; Birulés, I.; Verdaguer-Rodríguez, M.; López-Carrilero, R.; Ferrer-Quintero, M.; García-Mieres, H.; Díaz-Cutraro, L.; Grasa, E.; Pousa, E.; Lorente, E.; et al. Influence of Menstrual Cycle Length and Age at Menarche on Symptoms, Cognition, Social Cognition, and Metacognition in Patients with First-Episode Psychosis. *Women* **2022**, *2*, 135–146. [CrossRef]

74. Shen, Y.; Varma, D.S.; Zheng, Y.; Boc, J.; Hu, H. Age at Menarche and Depression: Results from the NHANES 2005-2016. *PeerJ* **2019**, *7*, e7150. [CrossRef] [PubMed]
75. Abraham, S.; Boyd, C.; Lal, M.; Luscombe, G.; Taylor, A. Time since Menarche, Weight Gain and Body Image Awareness among Adolescent Girls: Onset of Eating Disorders? *J. Psychosom. Obstet. Gynecol.* **2009**, *30*, 89–94. [CrossRef]
76. Steiner, M.; Dunn, E.; Born, L. Hormones and Mood: From Menarche to Menopause and Beyond. *J. Affect. Disord.* **2003**, *74*, 67–83. [CrossRef]
77. Chung, Y.S.; Poppe, A.; Novotny, S.; Epperson, C.N.; Kober, H.; Granger, D.A.; Blumberg, H.P.; Ochsner, K.; Gross, J.J.; Pearlson, G.; et al. A Preliminary Study of Association between Adolescent Estradiol Level and Dorsolateral Prefrontal Cortex Activity during Emotion Regulation. *Psychoneuroendocrinology* **2019**, *109*, 104398. [CrossRef]
78. Smith, K.E.; Mason, T.B.; Anderson, N.L.; Lavender, J.M. Unpacking Cognitive Emotion Regulation in Eating Disorder Psychopathology: The Differential Relationships between Rumination, Thought Suppression, and Eating Disorder Symptoms among Men and Women. *Eat. Behav.* **2019**, *32*, 95–100. [CrossRef]
79. Mikhail, M.E.; Kring, A.M. Emotion Regulation Strategy Use and Eating Disorder Symptoms in Daily Life. *Eat. Behav.* **2019**, *34*, 101315. [CrossRef]
80. Monteleone, A.M.; Cascino, G.; Marciello, F.; Abbate-Daga, G.; Baiano, M.; Balestrieri, M.; Barone, E.; Bertelli, S.; Carpiniello, B.; Castellini, G.; et al. Risk and Resilience Factors for Specific and General Psychopathology Worsening in People with Eating Disorders during COVID-19 Pandemic: A Retrospective Italian Multicentre Study. *Eat. Weight. Disord.* **2021**, *26*, 2443–2452. [CrossRef]
81. Meneguzzo, P.; Sala, A.; Merlino, L.; Ceccato, E.; Santonastaso, P. One Year of COVID-19 Pandemic on Patients with Eating Disorders, Healthy Sisters, and Community Women: Evidence of Psychological Vulnerabilities. *Eat. Weight. Disord. Stud. Anorex. Bulim. Obes.* **2022**, *27*, 3429–3438. [CrossRef]
82. Meneguzzo, P.; de Mico, A.; Gori, P.; Ajello, A.; Ceccato, E.; Consolati, M.D.; Vita, A.; Sala, A.; Santonastaso, P. A Multicenter Network Analysis Examining the Psychological Effects of the COVID-19 Pandemic on Patients with Eating Disorders versus Their Healthy Siblings. *J. Clin. Med.* **2022**, *11*, 7187. [CrossRef] [PubMed]
83. Masiero, M.; Mazzocco, K.; Harnois, C.; Cropley, M.; Pravettoni, G. From Individual To Social Trauma: Sources Of Everyday Trauma In Italy, The US And UK During The COVID-19 Pandemic. *J. Trauma Dissociation* **2020**, *21*, 513–519. [CrossRef] [PubMed]
84. Paslakis, G.; Scholz-Hehn, A.D.; Sommer, L.M.; Kühn, S. Implicit Bias to Food and Body Cues in Eating Disorders: A Systematic Review. *Eat. Weight. Disord.* **2021**, *26*, 1303–1321. [CrossRef] [PubMed]
85. Turgon, R.; Ruffault, A.; Juneau, C.; Blatier, C.; Shankland, R. Eating Disorder Treatment: A Systematic Review and Meta-Analysis of the Efficacy of Mindfulness-Based Programs. *Mindfulness* **2019**, *10*, 2225–2244. [CrossRef]

nutrients

MDPI

Article

Impact of COVID-19 Confinement on the Health-Related Habits of People at High Risk of Type 2 Diabetes

Darío Ochoa Esteban [1], Carmen Martin-Ridaura [1], Carmen Berlinches-Zapero [1], Dolores Ruiz-Fernández [1], Vanessa Sanz-Martín [1], Rosario Gavira-Izquierdo [1], Aitana Muñoz-Haba [2], Sebastià March [2] and Mercedes Ceinos-Arcones [1,*] on behalf of The ALAS Project Working Group

[1] Madrid Salud, Madrid City Council, 28007 Madrid, Spain
[2] Cooperativa APLICA, 28014 Madrid, Spain
* Correspondence: ceinosamm@madrid.es

Abstract: The general lockdown decreed in Spain due to the COVID-19 pandemic interrupted the ALAS health promotion intervention aimed at the population at high risk of suffering from type 2 diabetes. We conducted a descriptive study in 2020 through a telephone survey and a comparison with baseline data to determine the impact of confinement on the lifestyles of the participants. We collected sociodemographic variables and conducted assessments before/after confinement on general health status and lifestyle (sleep, physical activity and diet). Additionally, weight, BMI and adherence to a Mediterranean diet were assessed. Descriptive statistical analyses, comparisons of pre–post confinement data and logistic regression were carried out. A total of 387 individuals responded. Among them, 31.8% reported a worse perception of health after confinement, and 63.1% reported no change. Regarding exercise, 61.1% reduced their weekly physical activity time. Regarding diet, 34.4% perceived worse quality, and 53.4% reported no change, despite the fact that 89.4% declared changes in their eating practices. Weight and BMI decreased by 3.1%, and adherence to the Mediterranean diet improved from baseline. Confinement had a negative impact on the general health, diet, sleep and physical activity of this population (at risk of diabetes); however, weight and BMI decreased, and adherence to a Mediterranean diet improved.

Keywords: diabetes type 2; prediabetes; health promotion; general lockdown; COVID-19; lifestyle; diet; physical activity; obesity; Mediterranean diet; eating habits

1. Introduction

In March 2020, the WHO declared COVID-19 a global pandemic [1]. Following the increase in SARS-CoV-2 infections and observing the responses of many other countries, on 15 March 2020, a state of emergency was decreed in Spain [2]; the decree included general home confinement, which affected the majority of the population and sought to reduce the spread of the pandemic. In Spain, the confinement lasted until June, almost 100 days. It involved drastic changes in the day-to-day lives of the population; therefore, health habits were affected, as various studies have already reported [3–6]. There were significant impacts on the mental health of the population [7,8]; physical activity (PA) decreased [9]; sleep time and quality were altered [10]; and eating habits changed [11–13].

These impacts of general confinement applied at the population level have had important effects on society [14]. The effects that confinement may have on people with chronic diseases that are associated with lifestyle habits, for example, diabetes, or risk factors for developing chronic diseases are particularly relevant [15,16].

The high-risk intervention implemented through the ALAS programme includes individualised care and an intensive and structured group education workshop aimed at people with grade II overweight, obesity and/or a high risk of type 2 diabetes. The aim of the programme is to address and prevent these diseases by promoting changes in lifestyles, such as healthy eating, following a Mediterranean Diet and regular PA.

The programme began in 2011 and continues today. An effectiveness study was conducted with 1629 people who participated between 2016 and 2019 [17], concluding that (1) at the end of the intervention, 85% had lost weight, with 43% losing more than 5% of their baseline weight; (2) 22.3% of the people with obesity no longer presented obesity; (3) 35.1% of people classified as prediabetic according to the criteria of the American Diabetes Association became normoglycaemic; and (4) the effects on weight lasted at least 6 months after the end of the intervention.

Since September 2019, prior to confinement, 533 people have enrolled in the programme. Due to social isolation measures, the programme was interrupted. The participants who had started the intervention and the workshops were confined to their homes.

The purpose of this study is to describe the impact of confinement on the lifestyles of this sample of people with excess weight and/or a high risk of diabetes who had initiated an intervention programme to modify their health status.

2. Materials and Methods

2.1. Design

This was a descriptive study conducted in Madrid from May to June 2020, when the general confinement measures were de-escalated. A questionnaire was administered by telephone to people who were implementing the intervention programme. The interviews were conducted by health professionals affiliated with the programme and served simultaneously to maintain contact with the participants. Some data collected (weight, BMI, adherence to a Mediterranean diet) could be compared with baseline data collected previously for the programme. Therefore, a pre–post quasi-experimental design was possible.

2.2. Participants

The target population was the 553 people who had started the workshop in the ALAS programme between September 2019 and March 2020. The inclusion criteria to enter the programme were to be over 18 years of age and have one of the following: a BMI over 30 or BMI between 27 and 30 with an abdominal circumference risk (greater than 88 cm in women or 102 cm in men) or a FINDRISC (Finnish type 2 Diabetes Risk Score) greater than 14, this instrument having been validated for the Spanish context [18]. Participants accessed the programme through municipal health centres or were referred by municipal occupational health services (aimed at city council workers).

2.3. Description of the Programme

The ALAS intervention, based on the Diabetes Prevention Programme [19], consists of an intensive and structured intervention lasting 6 months that includes an individual intervention programme and a group education workshop [20] of 10 sessions (2 h each), over a period of 6 months, aimed at reducing weight, improving diet and increasing PA. The specific objectives of the programme include a 5% reduction in body weight, improvement in adherence to a Mediterranean diet, improvement in participation in PA and improvement in glycaemic status (for those with prediabetes).

2.4. Variables

The questionnaire, which combined ad hoc-designed impact measurements with validated tools, had different blocks:

2.4.1. Sociodemographic and Participation Data

Age, gender, country of birth, marital status, education level and employment status were collected. Information was also collected on housing and cohabitants during confinement: number of cohabitants, cohabitants under 14 years of age (yes/no), cohabitants older than 65 years of age or dependents (yes/no), size of residence (square metres) and access to the internet from home (Yes/No). With these variables, the density of inhabitants per household was calculated in square metres per inhabitant.

The following variable was collected for participation in the programme: number of sessions until interruption.

2.4.2. Health Status and Lifestyle Data

General health status: General health status before and after confinement was assessed through a Likert-type scale with 5 response options: very good, good, fair, poor and very poor.

Health-related life habits: Diet before and after confinement was assessed using a Likert-type scale with 5 response options (from 1 (unhealthy) to 5 (very healthy)). Participants were asked about changes in 12 practices related to diet that may have occurred due to confinement (better/same/worse): quantity of food consumed; variety of foods; regularity of meal times; snacking between meals; consumption of fresh, packaged and processed foods, sweets, pastries, soft drinks and alcoholic beverages; menu planning; time dedicated to cooking; and financial ability to implement a healthy diet. A validated questionnaire on adherence to a Mediterranean diet (MEDAS 14) [21] is used as an assessment tool for the programme. The questionnaire has 14 items related to the weekly frequency of the consumption of certain products, such as vegetables, oil and legumes. Each item has a cut-off value based on recommendations, and the global scale differentiates among high, medium and low adherence to a Mediterranean diet. Regarding physical exercise, the number of days per week and the average time spent exercising before and after confinement were assessed. Sleep time (total, day and night) and sleep quality (reduced or maintained) during confinement were also investigated.

2.4.3. Anthropometric Data

While baseline data were collected by programme professionals taking actual measurements, weight and height during confinement were self-reported, from which the BMI was calculated.

These data were paired to baseline data (weight, BMI and diet adherence) collected through the programme at the beginning of the intervention while maintaining anonymity. Variations in weight, BMI and diet adherence before and after confinement were calculated. In addition, the percentage of people who had achieved a 5% reduction in baseline weight was calculated; this variable is one of the main evaluation objectives of the ALAS programme.

2.5. *Analysis*

A descriptive analysis of all data was performed. The health and lifestyle variables were crossed, and chi-square tests and ANOVA were used to explore bivariate relationships. The factors related to the deterioration of general health or diet were adjusted using logistic regression models; the adjusted ORs are presented with the corresponding confidence intervals. The models were adjusted for all of the variables that had a significance of at least 90% and whose omission did not alter the parameters of the remaining variables by more than 5%.

For the pre–post analysis, weight, BMI, % obesity and adherence to a Mediterranean diet were compared using t-tests. Logistic regression analysis was performed to adjust the factors related to achieving a 5% reduction in baseline weight.

2.6. *Ethical Aspects*

The study complied with the principles of the Declaration of Helsinki of 2013. All participants gave their informed consent to answer the questionnaire and were informed that their collaboration was voluntary, anonymous and not conditional on their subsequent participation in the programme. The study was approved by and received ethical consent from the municipal health promotion service of Madrid Salud, which ensured anonymity throughout the process.

3. Results

3.1. Description of the Sample

A total of 387 (72.6%) of the 533 participants enrolled in the ALAS programme from September 2019 to February 2020 responded to the survey. There were no differences between the participants and non-participants in terms of age and sex, although the participants had a higher percentage of people with obesity at baseline (69.4 vs. 59.1) and therefore higher weight, waist circumference and baseline score of FINDRISK.

The mean age of the participants was 57.5 years (SD = 12.1; Rank = 18 − 86), and 77.9% of the sample were women. A total of 25.6% had not participated in any workshop sessions, 42% had participated in 1 to 5 sessions, and 32.4% had participated in 6 or more (complete intervention according to the ALAS programme protocol). The sociodemographic and housing characteristics of the sample are shown in Table 1, stratified by gender.

Table 1. Sociodemographic and housing characteristics of the sample by gender.

	Total N (% Column)	Men N (% Column)	Women N (% Column)	*p* *
Sample	387	85	299	-
		Sociodemographic variables		
Age				
≤44 years	52 (13.7)	13 (15.3)	39 (13.4)	0.932
45–54 years	81 (21.4)	18 (21.2)	63 (21.6)	
55–64 years	134 (35.4)	31 (36.5)	102 (39.4)	
≥65 years	112 (29.6)	23 (27.1)	88 (30.1)	
Origin				
Spain	349 (90.4)	80 (94.1)	267 (89.3)	0.498
Latin America	32 (8.3)	5 (5.9)	27 (9)	
Other	5 (1.3)	-	5 (1.7)	
Marital status				
Married or with a partner	219 (57.6)	61 (74.4)	157 (53.8)	0.001
Single	74 (19.5)	16 (19.5)	58 (19.9)	
Separated or divorced	43 (11.3)	2 (2.4)	40 (13.7)	
Widowed	40 (10.5)	3 (3.7)	37 (12.7)	
Complete studies				
Primary or lower	128 (34.6)	24 (29.3)	102 (35.7)	0.492
Secondary	45 (12.2)	13 (15.9)	32 (11.2)	
Baccalaureate or VE	92 (24.9)	19 (23.2)	73 (25.5)	
University	105 (28.4)	26 (31.7)	79 (27.6)	
Employment situation				
Working	153 (41.2)	38 (46.9)	115 (39.9)	0.038
Unemployed	47 (12.7)	5 (6.2)	42 (14.6)	
Retired/pension	146 (39.4)	37 (45.7)	107 (37.2)	
Home care	23 (6.2)	1 (1.2)	22 (7.6)	
Student	2 (0.5)	-	2 (0.7)	
		Housing variables		
Number of cohabitants				
One	81 (21.8)	9 (11)	71 (24.7)	0.038
Two	151 (40.7)	41 (50)	109 (38)	
Three	73 (19.7)	15 (18.3)	58 (20.2)	
More than three	66 (17.8)	17 (20.7)	49 (17.1)	

Table 1. *Cont.*

	Total N (% Column)	Men N (% Column)	Women N (% Column)	p *
Household size (m^2)				
≤60	73 (20.6)	17 (22.1)	55 (20)	0.014
61–80	136 (38.4)	26 (33.8)	109 (39.6)	
81–100	84 (23.7)	12 (15.6)	72 (26.2)	
>100	61 (17.2)	22 (28.6)	39 (14.2)	
Density of household inhabitants (m^{2}x cohabitant)				
≤25	91 (25.8)	19 (25)	72 (26.2)	0.972
25.5–37.5	95 (26.9)	22 (28.9)	72 (26.2)	
38–50	66 (18.7)	14 (18.4)	52 (18.9)	
>50	101 (28.6)	21 (27.6)	79 (28.7)	
Cohabitants in charge				
Children	52 (14)	12 (14.5)	39 (13.6)	0.849
Seniors/dependents	45 (12.2)	8 (9.8)	37 (13)	0.432
With internet access	319 (86.7)	73 (88)	244 (86.2)	0.684

* *p* value for the chi-square test of the relationship between the variable and gender, where <0.05 is considered a significant relationship.

3.2. Impact on Self-Perceived Health

Before confinement, 73.9% of the participants claimed to have good or very good health, with 19.4% and 6.7% reporting fair and bad or very bad health, respectively; during confinement, these percentages changed to 53.2%, 33.9% and 12.9%, respectively. A total of 31.8% of the participants reported a worse perception of their health after confinement, with 63.1% maintaining their self-perception of health and 5.1% reporting an increase. The worsening of the self-perceived state of health after the onset of the pandemic was related to being affected by obesity, being single, separated or divorced, and living with dependents; in contrast, living with minors was associated with preventing a deterioration in health (Table 2). When these factors were adjusted in a logistic regression model that included age and self-perceived health status before the pandemic, the effect of living with minors as a protective factor against deterioration was maintained (OR = 0.16; 95% CI = 0.06−0.43), and obesity was maintained as a risk factor (OR = 2.41; 95% CI = 1.36−4.28).

Table 2. Factors related to the deterioration of self-perceived health during confinement.

	Worse Health Status N (% Row)	Not Worse Health Status N (% Row)	p *
Sample	118 (31.8)	253 (68.2)	
Age			
≤44 years	17 (37.8)	28 (62.2)	0.117
45–54 years	32 (41)	46 (59)	
55–64 years	41 (31.5)	89 (68.5)	
≥65 years	28 (25.2)	83 (74.8)	
Gender			
Men	27 (32.5)	56 (67.5)	0.903
Women	91 (31.8)	195 (68.2)	
Pre-confinement BMI classification			
Normal weight	4 (22.2)	14 (77.8)	0.004
Overweight	19 (22.1)	67 (77.9)	
Obesity	94 (40.9)	136 (59.1)	
Origin			
Spain	104 (30.9)	233 (69.1)	0.218
Other	14 (41.2)	20 (58.8)	

Table 2. *Cont.*

	Worse Health Status N (% Row)	Not Worse Health Status N (% Row)	*p* *
Marital status			
Married or with a partner	60 (28.2)	153 (71.8)	
Single	18 (41.9)	25 (58.1)	0.020
Separated or divorced	30 (42.9)	40 (57.1)	
Widowed	8 (20)	32 (80)	
Complete studies			
Primary or lower	31 (25.2)	92 (74.8)	
Secondary	13 (29.5)	31 (70.5)	0.129
Baccalaureate or VE	34 (38.2)	55 (61.8)	
University	39 (37.5)	65 (62.5)	
Employment situation			
Working	52 (35.1)	96 (64.9)	
Unemployed	15 (34.1)	29 (65.9)	0.559
Retired/pension	43 (29.5)	103 (70.5)	
Home care	5 (22.7)	17 (77.3)	
Access through occupational health			
Yes	14 (29.2)	34 (70.8)	0.674
No	104 (32.2)	219 (67.8)	
Number of cohabitants			
One	30 (37)	51 (63)	0.219
More than one	86 (29.9)	202 (70.1)	
Household size (m^2)			
≤60	25 (34.7)	47 (65.3)	
61–80	38 (27.9)	98 (72.1)	0.325
81–100	33 (39.3)	51 (60.7)	
>100	18 (29.5)	43 (70.5)	
Density of household inhabitants (m^2x cohabitant)			
≤25	34 (37.8)	56 (62.2)	
25.5–37.5	31 (32.6)	64 (67.4)	0.259
38–50	15 (22.7)	51 (77.3)	
>50	33 (32.7)	68 (67.3)	
Cohabiting minors			
Yes	5 (9.8)	46 (90.2)	0.000
No	112 (35.1)	207 (64.9)	
Cohabiting elderly individuals/dependents			
Yes	21 (46.7)	24 (53.3)	0.022
No	96 (29.7)	227 (70.3)	
Internet access			
Yes	107 (33.6)	211 (66.4)	0.118
No	11 (22.4)	38 (77.6)	

* *p* value for the chi-square test of the relationship between the variable and worsening health status, where <0.05 is considered a significant relationship.

3.3. Impact on Sleep Habits

A total of 28.3% of the participants reduced their total sleep time during confinement, 59.3% maintained it, and 12.4% increased it. Regarding sleep quality, 39.6% believed that it worsened during confinement; for 53.9%, sleep quality remained the same, and for 6.5%, it improved. Sleep time at night decreased for 35% of the people surveyed, was similar to that before confinement for 55.5% and increased for 9.4%. Simultaneously, sleep time during daylight hours increased for 15.4% of the people surveyed, remained the same for 75.7% and decreased for 8.9%.

The worsening of sleep quality was related to being a woman or being unemployed ($p < 0.05$). Sleeping less during the day was statistically significantly related to having children or living with older dependents. The worsening of general health and diet during confinement was also significantly related to a worse quantity and quality of sleep. Age was also a factor. Younger people had greater sleep disturbances and worse sleep time and quality than did older people: 51.1% of participants under 45 years of age perceived worse sleep quality during confinement, followed by those who were 45–54 years of age, 43%; between 55 and 64 years of age, 40.3%; and those over 65 years of age, 31.5%.

3.4. Impact on Physical Exercise

Before confinement, 8.1% of the participants did not perform any type of weekly physical exercise; after confinement, this percentage increased to 28.6%, while 61% decreased the time they dedicated to physical exercise, with 24.4% reporting an increase and 14.6% reporting no change. Participants were physically active, on average, for 5 h per week (range: 0–35; SD = 4.2) before confinement and 3.4 h on average after confinement (range: 0–35; SD = 4). Before confinement, 60.3% met the recommendation of 30 min of physical exercise daily; after confinement, this percentage decreased to 40.8%.

The impact of confinement on PA was related to previous participation in PA (Table 3). A reduction in PA was related to greater-than-average participation in PA previously. In contrast, an increase in PA was related to participating in less PA before confinement. The participants who reported an increase in PA did so for an average of 3.9 h per week (SD = 4.3), 173% (SD = 214.5) of their pre-pandemic PA time. Those who reduced their participation in PA did so by an average of 4.2 h per week (SD = 3.6), 72% of the pre-pandemic PA time (SD = 28.7). These variations were significant (t-test, $p < 0.01$). Among the people who increased their PA, 36.7% met the recommendation of 30 min per day before confinement, increasing to 84.4% after confinement; for those who reduced their PA, compliance decreased from 73.8% to 22.2%.

Table 3. Relationship of the impact of confinement on physical activity based on previous activity.

	All	Increase PA	Maintain PA	Reduce PA	*p*
(% row)	100	24.4	14.6	61.0	-
Weekly PA time PRE (hours x week)	5.0	2.9	3.9	6.1	<0.01 *
Weekly PA time POST (hours x week)	3.4	6.8	3.9	1.8	<0.01 *
Variation in weekly PA time PRE–POST (hours x week)	1.6	3.9	-	−4.2	<0.01 *
Mean % of PA time variation (% of PRE time)	−8.7	173	-	−72.6	<0.01 *

PA = physical activity. * ANOVA of the time dedicated to PA according to impact (increase/maintain/decrease).

3.5. Impact on Diet

The average assessment of eating practised before confinement was 3.78 (SD = 0.89) on a scale of 1 to 5 (1: unhealthy; 5: very healthy). After confinement, the average was 3.47 (SD = 1.09), representing a significant difference (paired-samples Student's t-test, $p < 0.001$). A total of 34% reported a worse diet (scored worse) after confinement, 53.4% maintained their diet without changes, and 12.2% improved their diet. The relationship between the characteristics of the sample and the impact of confinement on the self-perception of their diet is shown in Table 4, and the adjustment by logistic regression is shown in Table 5. The negative impact of confinement on diet, adjusted for the assessment of diet before confinement (those who had a better diet experienced a greater negative impact), was significantly related to being middle-aged (45–64 years), living in a residence with less than 25 m^2 per cohabitant, living with elderly individuals or dependent people and being affected by obesity.

Table 4. Relationship between factors and the impact of confinement on diet.

	Worse Diet N (% Row)	Maintained Diet N (% Row)	Improved Diet N (% Row)	*p* *
Sample	127 (34.4)	197 (53.4)	45 (12.2)	
Age				
≤44 years	16 (35.6)	20 (44.4)	9 (20)	
45–54 years	35 (44.9)	31 (39.7)	12 (15.4)	
55–64 years	54 (41.9)	63 (48.8)	12 (9.3)	0.000
≥65 years	20 (18.2)	79 (71.8)	11 (10)	
Gender				
Men	27 (32.9)	45 (54.9)	10 (12.2)	
Women	100 (35.1)	150 (52.6)	35 (12.3)	0.929
Pre-confinement BMI classification				
Normal weight	4 (22.2)	12 (66.7)	2 (11.1)	
Overweight	19 (22)	55 (64)	12 (14)	0.023
Obesity	94 (40.9)	110 (47.8)	26 (11.3)	
Origin				
Spain	109 (32.6)	181 (54.2)	44 (13.2)	
Other	18 (51.4)	16 (45.7)	1 (2.9)	0.040
Marital status				
Married or with a partner	75 (35.7)	115 (54.8)	20 (9.5)	
Single	19 (44.2)	20 (46.5)	4 (9.3)	
Separated or divorced	25 (35.2)	31 (43.7)	15 (21.1)	0.012
Widowed	6 (15)	29 (72.5)	5 (12.5)	
Complete studies				
Primary or lower	34 (27.6)	74 (60.2)	15 (12.2)	
Secondary	17 (37.8)	27 (60)	1 (2.2)	
Baccalaureate or VE	35 (39.8)	46 (52.3)	7 (8)	0.012
University	37 (35.9)	45 (43.7)	21 (20.4)	
Employment situation				
Working	62 (41.9)	65 (43.9)	21 (14.3)	
Unemployed	21 (46.7)	21 (46.7)	3 (6.7)	
Retired/pension	35 (24.3)	92 (63.9)	17 (11.8)	0.006
Home care	6 (28.6)	14 (66.7)	1 (4.8)	
Access through occupational health				
Yes	20 (41.7)	22 (45.8)	6 (12.5)	
No	107 (33.3)	175 (54.5)	39 (12.1)	0.488
Number of cohabitants				
One	21 (37.1)	147 (51.4)	33 (11.5)	
More than one	106 (25.9)	48 (59.3)	12 (14.8)	0.170
Household size (m^2)				
≤60	22 (30.1)	44 (60.3)	7 (9.6)	
61–80	44 (32.6)	72 (53.3)	19 (14.1)	
81–100	34 (41)	38 (45.8)	11 (13.3)	0.653
>100	19 (31.7)	34 (56.7)	7 (11.7)	
Density of household inhabitants (m^{2}x cohabitant)				
≤25	48 (52.7)	37 (40.7)	6 (6.6)	
25.5–37.5	28 (29.5)	53 (55.8)	14 (14.7)	
38–50	18 (28.1)	36 (56.3)	10 (15.6)	0.002
>50	25 (25)	61 (61)	14 (14)	
Cohabiting minors				
Yes	107 (33.9)	173 (54.7)	36 (11.4)	
No	20 (38.5))	23 (44.2)	9 (17.3)	0.290
Elderly individuals or cohabiting dependents				
Yes	102 (31.7)	179 (55.6)	41 (12.7)	
No	24 (53.3)	17 (37.8)	4 (8.9)	0.016
Internet access				
Yes	9 (18.4)	34 (69.4)	6 (12.2)	
No	118 (37.3)	159 (50.3)	39 (12.3)	0.026

* *p* value for the chi-square test of the relationship between the variable and worsening health status, where <0.05 is considered a significant relationship.

Table 5. Adjusted effects of variables related to post-confinement diet deterioration.

	Adjusted OR	95% CI
Cohabiting with elderly individuals or dependents		
No	Ref	-
Yes	2.55	1.22–5.32
Age		
≤44 years	1.47	0.58–3.73
45–54 years	3.16	1.42–7.02
55–64 years	2.51	1.27–4.96
≥65 years	Ref	-
Density of household inhabitants		
≤25 m^2x cohabitant	2.70	1.26–5.78
25.5–37.5 m^2x cohabitant	1.10	0.53–2.31
38–50 m^2x cohabitant	1.68	0.74–3.78
>50 m^2x cohabitant	Ref	-
Pre-confinement BMI classification		
Normal/overweight	Ref	-
Obesity	2.24	1.19–4.21
Assessment of diet BEFORE confinement (scale 1–5)	1.54	1.12–2.12

Regarding the 12 dietary practices that were assessed (Table 6), 89.4% of the participants had modified at least 1 during confinement (average of 4.3 practices changed). Modifying a practice (increasing or decreasing) was related to a self-perception of the deterioration of diet during the pandemic: 37.3% of those who modified their practices reported a worse diet, and among those who did not change dietary practices, 10.3% reported a worse diet.

Table 6. Change in dietary practices during confinement.

	Change in Practices during Confinement		
	Less **N (% Row)**	**Same** **N (% Row)**	**Greater** **N (% Row)**
Healthy practices			
Time spent cooking	24 (6.5)	217 (58.6)	129 (34.9)
Planning menus	61 (16.5)	244 (65.9)	65 (17.6)
Consumption of fresh food	43 (11.6)	267 (72.2)	60 (16.2)
Variety of foods	64 (17.3)	257 (69.5)	49 (13.2)
Regularity of schedules	66 (17.8)	270 (73.0)	34 (9.2)
Financial ability to have a healthy diet	35 (9.5)	325 (87.8)	10 (2.7)
Unhealthy practices			
Unhealthy snacking between meals	55 (14.9)	150 (40.7)	164 (44.4)
Consumption of sweets and pastries	64 (17.3)	167 (45.3)	138 (38.4)
Amount of food consumed	64 (17.3)	173 (46.6)	134 (36.1)
Consumption of soft drinks	72 (19.5)	251 (67.5)	47 (12.7)
Consumption of packaged/processed foods	76 (20.5)	252 (68.1)	42 (11.4)
Consumption of alcoholic beverages	67 (18.1)	264 (71.4)	39 (10.5)

People over 65 years of age changed their practices the least, those younger than 45 years reduced healthy practices the most, and those aged 45–54 years increased their unhealthy practices the most. All participants who lived with minors changed their eating practices: this change was significantly associated ($p < 0.05$) with increasing healthy practices and reducing unhealthy practices, with the exception of the financial ability to implement a healthy diet, which was significantly lower in households with minors than in the rest (19.6% vs. 7.9%).

Some healthy practices increased, such as spending more time in the kitchen and the consumption of fresh food, and others were reduced, such as the regularity of meal times or the variety of foods consumed. The same happened with less healthy practices: snacking between meals, the consumption of sweets and pastries, and the amount of food consumed increased, and the consumption of packaged and processed foods, soft drinks and alcoholic beverages was reduced.

3.6. Weight Variation, BMI and Adherence to a Mediterranean Diet

Complete pre- and post-confinement weight data were obtained for 84% of the sample (n = 325). The majority (64.3%) lost at least 1 kg of weight from the baseline measurement. A total of 16.9% maintained their weight, and for 18.8%, their weight increased. On average, the surveyed population lost 2.7 kg (SD = 4.9), i.e., 3.1% of their baseline weight (SD = 5.5). Among the people who gained weight, the average gain was 3.4 kg (SD = 3.1) or 4% of their baseline weight (SD = 4), and among those who lost weight, the average loss was 5.2 kg (SD = 4) or 6% of their baseline weight (SD = 4.1). Weight loss was related to having accessed the programme through municipal occupational health services and being initially classified as overweight (not presenting obesity). Weight loss was also inversely related to the consumption of processed foods: among the people who reduced this consumption, 89.9% lost weight; among those who maintained consumption, 60.7% lost weight; and among those who increased consumption, 55.3% lost weight.

The variations in weight, BMI and adherence to a Mediterranean diet are shown in Table 7. Of the 322 people with complete data, 68.6% were classified as affected by obesity, 25.8% were overweight and 5.6% were normal weight at baseline; after confinement, the classification percentages were 58.1%, 34.8% and 7.1%, respectively. A total of 14% had a lower classification level (from obesity to overweight or from overweight to normal weight), 84.5% remained in the same classification, and 1.6% increased. Among the people initially classified as affected by obesity, 16.7% became overweight.

Table 7. Variations in weight, BMI and adherence to a Mediterranean diet.

	N	Pre	Post	Variation	p *
Weight (average kg)	325	86.9	84.2	2.7	<0.01
BMI	322	32.97	31.93	1.03	<0.01
% With obesity	322	68.6	58.1	10.6	<0.01
% High adherence to a Mediterranean diet	287	18.5	29.6	11.1	<0.01

* p value for the t-test of the measurement of the paired variable before and after confinement.

The variables that were statistically significantly related ($p < 0.05$) to the decrease in BMI after confinement were access to the programme through the municipal occupational health service (27.3% reduced BMI, compared to 11.9% of those who accessed the programme through community health centres), overcrowding in the home (17.3% in homes with less than 25 m^2/habitant, compared to 6.8% in those with more than 50 m^2/habitant) and employment status (17.7% in workers and 15.5% in retirees vs. 2.6% in unemployed and none of those who were dedicated to home care). The variation in BMI was not significantly related to the number of sessions performed, to the abandonment of the intervention or to not having performed any sessions.

Of the 287 people with complete MEDAS-14 data, 17.5% had low adherence to a Mediterranean diet at baseline, 65.9% had medium adherence, and 16.5% had high adherence. After confinement and starting the ALAS intervention, 6.1% had low adherence, 63.3% had medium adherence, and 30.6% had high adherence. A total of 31.4% improved their adherence, 56.4% maintained it, and 12.2% lowered it. A total of 83.3% of participants with low adherence at baseline improved.

This variation in adherence was not significantly related to the characteristics of the sample or to the variables related to the implementation of the intervention (number of sessions, dropouts, not performing any session), although the people who had participated in at least one session improved adherence more than those who had not participated in any (34.1% vs. 21.9%; $p = 0.064$).

Thirty-two percent (104) of the sample with complete weight data lost at least 5% of their initial weight, one of the main objectives of the ALAS programme. The adjusted factors related to this objective (Table 8) were accessing the programme through municipal occupational health services, having high adherence to a Mediterranean diet, having improved menu planning, being of non-Spanish origin and living in households with 80 to 100 m^2. Thirty-six percent of the people who completed three or more sessions of the ALAS workshop met the goal of a 5% or more reduction in their baseline weight, with 25.6% of those who did not participate in any sessions achieving the same goal; the difference was not statistically significant ($p = 0.401$ in the chi-square test).

Table 8. Adjusted effect of the factors related to a 5% reduction in baseline weight.

	Adjusted OR	95% CI
Access through occupational health		
No	Ref	-
Yes	4.93	2.09–11.61
Adherence to a Mediterranean diet		
Medium or low	Ref	
High	2.76	1.48–5.13
Planning menus		
Less than or equal	Ref	
Major	2.53	1.22–5.25
Origin		
Spain	Ref	
Other	3.28	1.25–8.64
Household size		
$\leq$ 60 m^2	2.15	0.75–6.13
61–80 m^2	1.47	0.57–3.77
81–100 m^2	3.13	1.19–8.27
> 100 m^2	Ref	
Density of household inhabitants		
$\leq$ 25 m^{2}x cohabitant	1.17	0.50–2.69
25.5–37.5 m^{2}x cohabitant	0.63	0.28–1.41
38–50 m^{2}x cohabitant	2.15	0.94–4.94
>50 m^{2}x cohabitant	Ref	

4. Discussion

The results of this study indicate the important impact that general confinement has had on the lifestyles of people at high risk of having diabetes who have begun an intensive health promotion intervention. Although the majority of people maintained their self-perception of health, approximately one-third reported worse sleep and diet quality and decreased PA. Changes in feeding practices varied and had positive and negative effects. In the sample studied, there was a significant reduction in weight and therefore in BMI, as well as an improvement in adherence to a Mediterranean diet.

Although 63.1% of the participants stated that there were no changes in their general self-perception of health during confinement, 31.8% reported that it was worse. This negative impact of confinement on the self-perception of health in general [22,23] and mental health specifically [24] has been noted in various studies conducted in recent years. There are factors, such as the context in which confinement took place [22] or the existence of chronic diseases or previous mental health problems [25], that have been related to this negative impact; however, they were not measured in our study. For other factors that have been shown to have a relationship, such as gender, age and social position [26], and that were collected here, the relationships were not maintained after adjustment. For our participants, being affected by obesity most conditioned the negative impact. While the relationship between obesity and poor health perception is known [27], there is little evidence about the relationship between obesity and the impact of confinement, although the media and scientific publications have already highlighted people with obesity as a potentially vulnerable population in the face of the imminence of the pandemic [28].

When adjusted for other factors, living with children (regardless of gender) emerged as a protective factor against the worsening of health or diet perceptions. This is an unexpected result; it is possible that confinement offered, among many difficulties, opportunities to improve care and dedicate more time to cohabitant minors. In fact, people with children were among those who had to make the most changes in their routines during confinement [29,30]. Other studies of confinement indicate that living with minors is related to a negative impact on diet [31]; however, the previous literature is divided as to whether it worsens or improves the household diet [32,33].

Our data indicate that the majority of participants maintained their usual sleep time, while 28% reduced their sleep time and 12% increased it. However, up to 40% reported worse sleep quality. A French study [34] and an Australian study [35] also found increases in the incidence of sleep problems compared to relevant evidence before confinement. Others [36,37] have also observed this increase, describing associated factors such as overexposure to blue light due to the increase in the use of devices and in teleworking.

Our results indicate a general decrease in the time dedicated to PA during confinement. This result is consistent with various studies in the general population conducted for the same period in Spain [38,39] and in approximately 20 studies in Italy reviewed by Zaccagni et al. [40]. The results also coincide with a Singaporean investigation based on data recorded automatically and continuously by bracelets [41]. In contrast, a few studies in European countries obtained contradictory results that indicate an increase in declared PA during confinement [42–44]. Consistent with our results, low compliance with the minimum recommended PA guidelines was also observed among the general population in Spain [39] and the United Kingdom [42].

Our data indicate that the decrease in PA was not homogeneous for all profiles: those who reported a higher level of PA before confinement experienced a strong decrease in compliance with the recommendation of 30 min of daily exercise, and for those who were less active pre-confinement, compliance with this standard significantly increased. This finding contrasts with the results of a study from Singapore [41]: i.e., people who decreased their PA the most were the most inactive before confinement, while those who were the most active pre-confinement maintained a profile with greater PA, despite a reduction.

Castañeda-Babarro [38] observed a similar reduction in PA that was more intense among originally more active individuals, a finding that could be explained by the forced closure of sports facilities during confinement and the lack of sports equipment at home. Additionally, among the most inactive people before confinement, moderate PA at home increased [43]. For the most inactive population, confinement represented an opportunity to dedicate time to PA, while those who were more active may have been limited by the inaccessibility of the spaces where they generally practised PA (gyms, swimming pools, parks, etc.).

Although half of the participants believed that there was no impact, one in three participants perceived that their diet had worsened with confinement. Several studies agree that diets were mostly maintained [5,43,45,46]. Some studies reported changes in eating habits; however, the direction of those changes is not clear [3,13]. A study by López-Moreno et al. [47] of the general population in Spain showed that 54% changed habits, with 38.4% improving and 16% worsening. Additionally, Net-Santé in France [31] indicated that 56.2% of participants changed their eating practices. As in our results, other studies agree that perceiving a deterioration in one's diet is related to age [48] and BMI [49]; however, these studies do not relate this perception to the conditions of the home or to living with elderly dependents.

A study conducted in various European countries by Skotnicka et al. [50] concluded that people with worse eating habits before confinement experienced a more negative impact on their diets. In contrast, in our study, the results indicate the opposite. Similar results were observed in a follow-up of US and UK cohorts [4].

Among the variability in changes in eating habits, evidence from European studies supports the results obtained in our study, i.e., increased time spent cooking [3,31,43,47,48]; the increased consumption of fresh products, fruits and vegetables [3,12,31,39,43,48]; and the increased consumption of snacks and sweets [3,12,31,47–51]. In our study, one-fifth of the participants reduced their consumption of ultra-processed foods, a finding that is consistent with some studies [43] but that also differs from other studies that warned of an increase [3,47,52]. In our study, 18.1% of the participants increased their consumption of alcohol, and 10.5% reduced their consumption; therefore, the impact is not clear. In other European studies, high variability is noted in the impact on alcohol consumption [3], with results that indicate that it increases [47–51] and others that indicate that it decreases [12,39,43,53]. These changes in habits seem to have produced slight improvements in adherence to a Mediterranean diet [50,53], which is considered heart-healthy. In our study, a significant increase in this adherence stands out (31.4%); however, the majority (56.4%) did not report significant changes.

These impacts on changes in habits (eating and PA) produced weight changes in people who lived in confinement; however, these include increases and decreases. Therefore, there is substantial variability [11,31,47,48], and at the population level, there does not seem to be a significant effect [54,55]. A longitudinal study in the United Kingdom reported that there was an increase in average weight at the beginning of confinement but that weight later decreased, resulting in no impact of confinement on weight [54].

There seems to be some consensus that the people who gained the most weight were those who were previously affected by obesity or overweight [3,11,54,56]. This contrasts with the results of our sample: 68.6% were initially affected by obesity, and the majority (64.3%) lost at least 1 kg of weight and therefore reduced their BMI.

A total of 32% of the sample achieved a reduction of 5% in their baseline weight, one of the main objectives of the ALAS programme. Although the design of this study did not allow the establishment of causal relationships, several hypotheses can be proposed to explain how there was a significant reduction in weight in this sample, while most studies in Spain and nearby countries indicate that, at the population level, there was no significant impact. The participants had decided to enrol in a health education programme to modify their lifestyle; therefore, they were people in a phase of preparation for action according to the transtheoretical model [57,58] and therefore more likely to make these changes. In addition, for this population, the ALAS programme serves as a motivator and supports the implementation of change. In fact, the 32% efficacy in the 5% reduction in baseline weight is similar to the 34% established in previous evaluations of the programme [17] for participants who did not complete the intervention. That is, the simple fact of having access to the intervention (without completing it) may promote change. For this study, intervention follow-up (number of sessions performed) did not have a significant effect. By entering the programme (an inclusion criterion in this study), participants meet for an individual session where they are weighed, measured, given questionnaires on lifestyle habits and provided information and brief advice to motivate change. This intervention and brief advice by themselves have already demonstrated some efficacy, although slight in magnitude, in promoting change [58–60]. Thus, although confinement seems to impact the weight of the population in a variable direction that results in modest changes at the population level, other factors were also at play in our study, such as the selection of a sample motivated to change and the fact that they were already enrolled in a health promotion programme. These last factors seem to have counteracted the effects of confinement described in other studies of the general population, particularly among people with obesity and overweight people.

Although the ALAS programme is carried out in municipal health centres, participants can either be recruited directly in these centres or start after a referral from the labour services of the city council. Those who enrol in the programme through the second path are mainly municipal officials, meaning they are civil servants, a socio-economic profile that is typically subjected to more favourable social determinants than the general population captured in universal access services. This may explain why the pre–post results of the intervention are better in the group of people who accessed the programme through an occupational health referral, as happened in the evaluation of the ALAS programme itself [17].

The study had a very high participation rate (72.6%), which is very valuable, given the difficult moment in which the questionnaire was administered. This success is attributed to the following: unlike most studies conducted during confinement, instead of opting for an online questionnaire, a telephone interview was conducted by professionals from their municipal health centres. Using this approach, barriers to the accessibility of online questionnaires were overcome; therefore, our sample has a higher proportion of elderly people and people with low levels of education than do other studies conducted in Spain in the same period [23,26,29,39,47,50,53]. Additionally, our sample is of particular interest for evaluating the impact of confinement because the participants were selected precisely because they are people with obesity and/or have a high risk of developing diabetes in the next 10 years if they do not modify their lifestyles. As in most studies carried out during confinement and referenced in this work, weight during confinement, as well as some of the impacts of confinement, was self-reported. This can potentially produce some biases; these should be similar to those that occur in the general population in other studies, and therefore, they should be comparable. There were statistically significant differences in the baseline weight of the participants compared to the non-participants, and a potential participation bias was checked. In this case, the results go against the hypothesis, because those who did not participate were precisely the ones who weighed less, so we understand that there was not a significant bias. Additionally, complete pre–post data on weight were not available for 62 participants (16% of the sample). Half of these cases were due to the fact that the baseline information could not be collected due to the start of the pandemic. The other half did not know their weight during confinement, as they may not have had the instruments to measure it. We believe that these data losses are not important enough to generate a bias in the study.

5. Conclusions

A significant percentage of people at risk of developing chronic diseases who enrolled in the ALAS health promotion programme between October 2019 and February 2020 perceived that the confinement experienced between March and June 2020 had a negative impact on their overall health, diet, sleep habits and PA. Interestingly, when compared with baseline data, weight and BMI decreased, and adherence to a Mediterranean diet improved. This positive effect on health could be explained by the participants being prepared to change and finding motivation and support through the ALAS programme. These results suggest that the development of public health promotion programmes with high-risk populations not only has a potential direct effect on their participants but can also protect them from the effects of unforeseen adverse situations threatening the stability of their health-related lifestyles, such as general confinement in this case. This could orient policymakers to continue promoting programmes targeting specific vulnerabilities and striving to sustain these interventions even in times of emergency. Our discussion also suggests that the satisfactory evolution of the participants during confinement might be related to the fact that the participants seem to be in an adequate phase of preparation for changes in their lifestyles. This leads to the recommendation that health promotion programmes adapt their interventions at the stage of change in the participants, i.e., raise motivation among those who do not present that attitude before prescribing specific routine transformations among those who are willing to introduce changes in their lifestyles. In any case, more research is needed to strengthen these ideas.

Author Contributions: Conceptualisation, D.O.E., C.M.-R., C.B.-Z., D.R.-F., V.S.-M., R.G.-I., A.M.-H., S.M. and M.C.-A.; data curation, D.O.E., C.M.-R., C.B.-Z., D.R.-F., V.S.-M. and R.G.-I.; formal analysis, S.M.; methodology, D.O.E., C.M.-R., C.B.-Z., D.R.-F., V.S.-M., R.G.-I., A.M.-H., S.M. and M.C.-A.; supervision, M.C.-A.; visualisation, S.M.; writing—original draft, A.M.-H. and S.M.; writing—review and editing, D.O.E., C.M.-R., C.B.-Z., D.R.-F., V.S.-M., R.G.-I., A.M.-H., S.M. and M.C.-A. All authors have read and agreed to the published version of the manuscript.

Funding: This study was funded by the Novo Nordisk group (CIF: A28081495), as part of the international project Cities Changing Diabetes.

Institutional Review Board Statement: This study was conducted in accordance with the Declaration of Helsinki and was approved by the municipal health promotion service of Madrid Salud.

Informed Consent Statement: Informed consent was obtained from all subjects involved in the study.

Data Availability Statement: The datasets used and/or analysed during the current study are available from the corresponding author on reasonable request.

Acknowledgments: We appreciate the collaboration of the ALAS programme participants in the study during a time of communication difficulties and social distancing.Additional Information: The ALAS Project Working Group (Affiliated to Madrid City Council, Madrid, Spain) members are: Dolores Rivas-Hernández, Saray Pino-Vega, Juan Manuel Melero-Rubio, Francisco Bordel-Nieto, Milagros Caballero-Jauregui, Isabel Corella-Monzon, Raquel Pino-Días, Carmen Cutanda-Rodriguez, Sergio Sánchez-Muñoz, José Manuel Fernández-Garrido, Carmen Morales-López, María Jesús Majarrez-Arias, Santiago Sancho, Nuria Calle Fernández, Mercedes Martínez-Cortes, Pilar García-Crespo, Carmen María León-Dominguez, Marina Pascual & Raquel Herrera.

Conflicts of Interest: The authors declare no conflict of interest. The funders had no role in the design of the study; in the collection, analyses or interpretation of data; in the writing of the manuscript; or in the decision to publish the results.

References

1. Cucinotta, D.; Vanelli, M. WHO declares COVID-19 a pandemic. *Acta Biomed.* **2020**, *91*, 157–160. [CrossRef]
2. *Real Decreto 463/2020, de 14 de Marzo, Por el Que se Declara el Estado de Alarma Para la Gestión de la Situación de Crisis Sanitaria Ocasionada Por El COVID-19*; Official Gazette of the Government of Spain (Boletín Oficial del Estado – BOE) nº67,(14-03-2020); Ministerio de la Presidencia, Relaciones con las Cortes y Memoria Democrática: Seville, Spain, 2020.
3. Doraiswamy, S.; Cheema, S.; Al Mulla, A.; Mamtani, R. COVID-19 lockdown and lifestyles: A narrative review. *F1000Research* **2021**, *10*, 363. [CrossRef]
4. Mazidii, M.; Leeming, E.; Merino, J.; Nguyen, L.; Selvachandran, S.; Maher, T.; Kadé, K.; Murray, B.; Graham, M.; Sudre, C.; et al. Impact of COVID-19 on Health Behaviours and Body Weight: A Prospective Observational Study in a Cohort of 1.1 Million UK and US Individuals. EuropePMC. 2021. Available online: https://www.researchsquare.com/article/rs-179013/v1 (accessed on 24 November 2022).
5. Yang, G.Y.; Lin, X.L.; Fang, A.P.; Zhu, H.L. Eating habits and lifestyles during the initial stage of the COVID-19 lockdown in China: A cROSS-sECTIONAL sTUDY. *Nutrients* **2021**, *13*, 970. [CrossRef]
6. Chiesa, V.; Antony, G.; Wismar, M.; Rechel, B. COVID-19 pandemic: Health impact of staying at home, social distancing and 'lockdown' measures-a systematic review of systematic reviews. *J. Public Health* **2021**, *43*, e462–e481. [CrossRef]
7. Nochaiwong, S.; Ruengorn, C.; Thavorn, K.; Hutton, B.; Awiphan, R.; Phosuya, C.; Ruanta, Y.; Wongpakaran, N.; Wongpakaran, T. Global prevalence of mental health issues among the general population during the coronavirus disease-2019 pandemic: A systematic review and meta-analysis. *Sci. Rep.* **2021**, *11*, 10173. [CrossRef]
8. Xiong, J.; Lipsitz, O.; Nasri, F.; Lui, L.M.W.; Gill, H.; Phan, L.; Chen-Li, D.; Iacobucci, M.; Ho, R.; Majeed, A.; et al. Impact of COVID-19 pandemic on mental health in the general population: A systematic review. *J. Affect. Disord.* **2020**, *277*, 55–64. [CrossRef]
9. Stockwell, S.; Trott, M.; Tully, M.; Shin, J.; Barnett, Y.; Butler, L.; McDermott, D.; Schuch, F.; Smith, L. Changes in physical activity and sedentary behaviours from before to during the COVID-19 pandemic lockdown: A systematic review. *BMJ Open Sport Exerc. Med.* **2021**, *7*, e000960. [CrossRef]
10. O'Regan, D.; Jackson, M.L.; Young, A.H.; Rosenzweig, I. Understanding the impact of the COVID-19 pandemic, lockdowns and social isolation on sleep quality. *Nat. Sci. Sleep* **2021**, *13*, 2053–2064. [CrossRef]

11. Khan, M.A.; Menon, P.; Govender, R.; Samra, A.M.A.; Allaham, K.K.; Nauman, J.; Östlundh, L.; Mustafa, H.; Smith, J.E.M.; AlKaabi, J.M. Systematic review of the effects of pandemic confinements on body weight and their determinants. *Br. J. Nutr.* **2022**, *127*, 298–317. [CrossRef]
12. Zupo, R.; Castellana, F.; Sardone, R.; Sila, A.; Giagulli, V.A.; Triggiani, V.; Cincione, R.I.; Giannelli, G.; De Pergola, G. Preliminary trajectories in dietary behaviors during the COVID-19 pandemic: A public health call to action to face obesity. *Int. J. Environ. Res. Public Health* **2020**, *17*, 7073. [CrossRef]
13. Bennett, G.; Young, E.; Butler, I.; Coe, S. The impact of lockdown during the COVID-19 outbreak on dietary habits in various population groups: A scoping review. *Front. Nutr.* **2021**, *8*, 626432. [CrossRef]
14. Rose, G. Sick individuals and sick populations. *Int. J. Epidemiol.* **1985**, *14*, 32–38. [CrossRef]
15. Zhang, X.; Devlin, H.M.; Smith, B.; Imperatore, G.; Thomas, W.; Lobelo, F.; Ali, M.K.; Norris, K.; Gruss, S.; Bardenheier, B.; et al. Effect of lifestyle interventions on cardiovascular risk factors among adults without impaired glucose tolerance or diabetes: A systematic review and meta-analysis. *PLoS ONE* **2017**, *12*, e0176436. [CrossRef]
16. Horton, E.S. Effects of lifestyle changes to reduce risks of diabetes and associated cardiovascular risks: Results from large scale efficacy trials. *Obesity* **2009**, *17* (Suppl. S3), S43–S48. [CrossRef]
17. Martin-Ridaura, C.; Ochoa-Esteban, D.; Berlinches-Zapero, C.; Ruiz-Fernández, D.; Sanz-Martín, V.; Gavira-Izquierdo, R.; March, S.; López-Toribio, M.; Ceinos-Arcones, M. Evaluation under real-life conditions of a lifestyle intervention for diabetes prevention developed by the municipal health services of Madrid, Spain. *Sci. Rep.* **2022**, *12*, 19700. [CrossRef]
18. Soriguer, F.; Valdes, S.; Tapia, M.J.; Esteva, I.; Ruiz de Adana, M.S.; Almaraz, M.C.; Morcillo, S.; García Fuentes, E.; Rodríguez, F.; Rojo-Martinez, G.; et al. Validation of the FINDRISC (FINnish Diabetes RIsk SCore) for prediction of the risk of type 2 diabetes in a population of southern Spain. *Pizarra Study. Med Clin.* **2012**, *138*, 371–376.
19. Knowler, W.C.; Barrett-Connor, E.; Fowler, S.E.; Hamman, R.F.; Lachin, J.M.; Walker, E.A.; Nathan, D.M. Reduction in the incidence of type 2 diabetes with lifestyle intervention or metformin. *N. Engl. J. Med.* **2002**, *346*, 393–403. [CrossRef]
20. Programa de Alimentación, Actividad Física y Salud (ALAS). Madrid Salud. Available online: https://madridsalud.es/programa-alas-alimentacion-actividad-fisica-y-salud/ (accessed on 24 November 2022).
21. Martínez-González, M.A.; García-Arellano, A.; Toledo, E.; Salas-Salvadó, J.; Buil-Cosiales, P.; Corella, D.; Covas, M.I.; Schröder, H.; Arós, F.; Gómez-Gracia, E.; et al. A 14-item Mediterranean diet assessment tool and obesity indexes among high-risk subjects: The PREDIMED trial. *PLoS ONE* **2012**, *7*, e43134. [CrossRef]
22. Le, K.; Nguyen, M. The psychological consequences of COVID-19 lockdowns. *Int. Rev. Appl. Econ.* **2021**, *35*, 147–163. [CrossRef]
23. Ripoll, J.; Contreras-Martos, S.; Esteva, M.; Soler, A.; Serrano-Ripoll, M.J. Mental health and psychological wellbeing during the COVID-19 lockdown: A longitudinal study in the Balearic Islands (Spain). *J. Clin. Med.* **2021**, *10*, 3191. [CrossRef]
24. Richter, D.; Riedel-Heller, S.; Zürcher, S.J. Mental health problems in the general population during and after the first lockdown phase due to the SARS-Cov-2 pandemic: Rapid review of multi-wave studies. *Epidemiol. Psychiatr. Sci.* **2021**, *30*, e27. [CrossRef]
25. Wang, Y.; Shi, L.; Que, J.; Lu, Q.; Liu, L.; Lu, Z.; Xu, Y.; Liu, J.; Sun, Y.; Meng, S.; et al. The impact of quarantine on mental health status among general population in China during the COVID-19 pandemic. *Mol. Psychiatry* **2021**, *26*, 4813–4822. [CrossRef]
26. Jacques-Aviñó, C.; López-Jiménez, T.; Medina-Perucha, L.; de Bont, J.; Gonçalves, A.Q.; Duarte-Salles, T.; Berenguera, A. Gender-based approach on the social impact and mental health in Spain during COVID-19 lockdown: A cross-sectional study. *BMJ Open* **2020**, *10*, e044617. [CrossRef]
27. Sayón-Orea, C.; Santiago, S.; Bes-Rastrollo, M.; Martínez-González, M.; Pastor, M.; Moreno-Aliaga, M.; Tur, J.; Garcia, A.; Martínez, J. Determinants of self-rated health perception in a sample of a physically active population: PLENUFAR VI study. *Int. J. Environ. Res. Public Health* **2018**, *15*, 2104. [CrossRef]
28. Katsoulis, M.; Pasea, L.; Lai, A.G.; Dobson, R.J.B.; Denaxas, S.; Hemingway, H.; Banerjee, A. Obesity during the COVID-19 pandemic: Both cause of high risk and potential effect of lockdown? A population-based electronic health record study. *Public Health* **2021**, *191*, 41–47. [CrossRef]
29. Navas-Martín, M.Á.; López-Bueno, J.A.; Oteiza, I.; Cuerdo-Vilches, T. Routines, time dedication and habit changes in spanish homes during the COVID-19 lockdown. A large cross-sectional survey. *Int. J. Environ. Res. Public Health* **2021**, *18*, 12176. [CrossRef] [PubMed]
30. Di Giorgio, E.; Di Riso, D.; Mioni, G.; Cellini, N. The interplay between mothers' and children behavioral and psychological factors during COVID-19: An Italian study. *Eur. Child Adolesc. Psychiatry* **2020**, *30*, 1401–1412. [CrossRef]
31. Deschasaux-Tanguy, M.; Druesne-Pecollo, N.; Esseddik, Y.; de Edelenyi, F.S.; Alles, B.; Andreeva, V.A.; Baudry, J.; Charreire, H.; Deschamps, V.; Egnell, M.; et al. Diet and physical activity during the coronavirus disease 2019 (COVID-19) lockdown (March-May 2020): Results from the French NutriNet-Sante cohort study. *Am. J. Clin. Nutr.* **2021**, *113*, 924–938. [CrossRef]
32. Roos, E.; Lahelma, E.; Virtanen, M.; Prättälä, R.; Pietinen, P. Gender, socioeconomic status and family status as determinants of food behaviour. *Soc. Sci. Med.* **1998**, *46*, 1519–1529. [CrossRef]
33. Berge, J.M.; Larson, N.; Bauer, K.W.; Neumark-Sztainer, D. Are parents of young children practicing healthy nutrition and physical activity behaviors? *Pediatrics* **2011**, *127*, 881–887. [CrossRef]

34. Beck, F.; Léger, D.; Fressard, L.; Peretti-Watel, P.; Verger, P. Covid-19 health crisis and lockdown associated with high level of sleep complaints and hypnotic uptake at the population level. *J. Sleep Res.* **2020**, *30*, e13119. [CrossRef]
35. Stanton, R.; To, Q.G.; Khalesi, S.; Williams, S.L.; Alley, S.J.; Thwaite, T.L.; Fenning, A.S.; Vandelanotte, C. Depression, anxiety and stress during COVID-19: Associations with changes in physical activity, sleep, tobacco and alcohol use in australian adults. *Int. J. Environ. Res. Public Health* **2020**, *17*, 4065. [CrossRef]
36. Marelli, S.; Castelnuovo, A.; Somma, A.; Castronovo, V.; Mombelli, S.; Bottoni, D.; Leitner, C.; Fossati, A.; Ferini-Strambi, L. Impact of COVID-19 lockdown on sleep quality in university students and administration staff. *J. Neurol.* **2020**, *268*, 8–15. [CrossRef] [PubMed]
37. Lahiri, A.; Jha, S.S.; Acharya, R.; Dey, A.; Chakraborty, A. Correlates of insomnia among the adults during COVID19 pandemic: Evidence from an online survey in India. *Sleep Med.* **2021**, *77*, 66–73. [CrossRef] [PubMed]
38. Castañeda-Babarro, A.; Arbillaga-Etxarri, A.; Gutiérrez-Santamaría, B.; Coca, A. Physical activity change during COVID-19 confinement. *Int. J. Environ. Res. Public Health* **2020**, *17*, 6878. [CrossRef]
39. López-Bueno, R.; Calatayud, J.; Casaña, J.; Casajús, J.A.; Smith, L.; Tully, M.A.; Andersen, L.L.; López-Sánchez, G.F. COVID-19 confinement and health risk behaviors in Spain. *Front. Psychol.* **2020**, *11*, 1426. [CrossRef]
40. Zaccagni, L.; Toselli, S.; Barbieri, D. Physical activity during COVID-19 lockdown in Italy: A systematic review. *Int. J. Environ. Res. Public Health* **2021**, *18*, 6416. [CrossRef]
41. Ong, J.L.; Lau, T.; Massar, S.A.A.; Chong, Z.T.; Ng, B.K.L.; Koek, D.; Zhao, W.; Yeo, B.T.T.; Cheong, K.; Chee, M.W.L. COVID-19-related mobility reduction: Heterogenous effects on sleep and physical activity rhythms. *Sleep* **2020**, *44*, zsaa179. [CrossRef]
42. Spence, J.C.; Rhodes, R.E.; McCurdy, A.; Mangan, A.; Hopkins, D.; Mummery, W.K. Determinants of physical activity among adults in the United Kingdom during the COVID-19 pandemic: The DUK-COVID study. *Br. J. Health Psychol.* **2020**, *26*, 588–605. [CrossRef]
43. Di Renzo, L.; Gualtieri, P.; Pivari, F.; Soldati, L.; Attinà, A.; Cinelli, G.; Leggeri, C.; Caparello, G.; Barrea, L.; Scerbo, F.; et al. Eating habits and lifestyle changes during COVID-19 lockdown: An Italian survey. *J. Transl. Med.* **2020**, *18*, 229. [CrossRef]
44. Di Corrado, D.; Magnano, P.; Muzii, B.; Coco, M.; Guarnera, M.; De Lucia, S.; Maldonato, N.M. Effects of social distancing on psychological state and physical activity routines during the COVID-19 pandemic. *Sport Sci. Health* **2020**, *16*, 619–624. [CrossRef]
45. Herle, M.; Smith, A.D.; Bu, F.; Steptoe, A.; Fancourt, D. Trajectories of eating behavior during COVID-19 lockdown: Longitudinal analyses of 22,374 adults. *Clin. Nutr. ESPEN* **2021**, *42*, 158–165. [CrossRef] [PubMed]
46. Romeo-Arroyo, E.; Mora, M.; Vázquez-Araújo, L. Consumer behavior in confinement times: Food choice and cooking attitudes in Spain. *Int. J. Gastron. Food Sci.* **2020**, *21*, 100226. [CrossRef]
47. López-Moreno, M.; López, M.T.I.; Miguel, M.; Garcés-Rimón, M. Physical and psychological effects related to food habits and lifestyle changes derived from COVID-19 home confinement in the Spanish population. *Nutrients* **2020**, *12*, 3445. [CrossRef] [PubMed]
48. Skotnicka, M.; Karwowska, K.; Kłobukowski, F.; Wasilewska, E.; Małgorzewicz, S. Dietary habits before and during the COVID-19 epidemic in selected European countries. *Nutrients* **2021**, *13*, 1690. [CrossRef] [PubMed]
49. Robinson, E.; Boyland, E.; Chisholm, A.; Harrold, J.; Maloney, N.G.; Marty, L.; Mead, B.R.; Noonan, R.; Hardman, C.A. Obesity, eating behavior and physical activity during COVID-19 lockdown: A study of UK adults. *Appetite* **2021**, *156*, 104853. [CrossRef]
50. Sánchez-Sánchez, E.; Ramírez-Vargas, G.; Avellaneda-López, Y.; Orellana-Pecino, J.I.; García-Marín, E.; Díaz-Jimenez, J. Eating habits and physical activity of the Spanish population during the COVID-19 pandemic period. *Nutrients* **2020**, *12*, 2826. [CrossRef]
51. Bakaloudi, D.R.; Jeyakumar, D.T.; Jayawardena, R.; Chourdakis, M. The impact of COVID-19 lockdown on snacking habits, fast-food and alcohol consumption: A systematic review of the evidence. *Clin. Nutr.* **2021**, *41*, 3038–3045. [CrossRef]
52. Bonaccio, M.; Costanzo, S.; Ruggiero, E.; Persichillo, M.; Esposito, S.; Olivieri, M.; Di Castelnuovo, A.; Cerletti, C.; Donati, M.B.; de Gaetano, G.; et al. Changes in ultra-processed food consumption during the first Italian lockdown following the COVID-19 pandemic and major correlates: Results from two population-based cohorts. *Public Health Nutr.* **2021**, *24*, 3905–3915. [CrossRef]
53. Rodríguez-Pérez, C.; Molina-Montes, E.; Verardo, V.; Artacho, R.; García-Villanova, B.; Guerra-Hernández, E.J.; Ruíz-López, M.D. Changes in dietary behaviours during the COVID-19 outbreak confinement in the Spanish COVIDiet study. *Nutrients* **2020**, *12*, 1730. [CrossRef]
54. Dicken, S.J.; Mitchell, J.J.; Le Vay, J.N.; Beard, E.; Kale, D.; Herbec, A.; Shahab, L. Impact of COVID-19 pandemic on weight and BMI among UK adults: A longitudinal analysis of data from the HEBECO study. *Nutrients* **2021**, *13*, 2911. [CrossRef] [PubMed]
55. Dicken, S.J.; Mitchell, J.J.; Newberry Le Vay, J.; Beard, E.; Kale, D.; Herbec, A.; Shahab, L. Impact of the COVID-19 pandemic on diet behaviour among UK adults: A longitudinal analysis of the HEBECO study. *Front. Nutr.* **2021**, *8*, 788043. [CrossRef]
56. Chew, H.S.J.; Lopez, V. Global impact of COVID-19 on weight and weight-related behaviors in the adult population: A scoping review. *Int. J. Environ. Res. Public Health* **2021**, *18*, 1876. [CrossRef] [PubMed]
57. Prochaska, J.O.; Velicer, W.F. The transtheoretical model of health behavior change. *Am. J. Health Promot.* **1997**, *12*, 38–48. [CrossRef] [PubMed]

58. Ezika, E. Transtheoretical model as a framework for promoting cardiovascular health through behaviour change: A systematic review. *Int. Res. J. Public Health* **2020**, *4*, 46. [CrossRef]
59. Vijay, G.; Wilson, E.C.F.; Suhrcke, M.; Hardeman, W.; Sutton, S. Are brief interventions to increase physical activity cost-effective? A systematic review. *Br. J. Sport. Med.* **2015**, *50*, 408–417. [CrossRef]
60. Zhang, A.; Franklin, C.; Currin-McCulloch, J.; Park, S.; Kim, J. The effectiveness of strength-based, solution-focused brief therapy in medical settings: A systematic review and meta-analysis of randomized controlled trials. *J. Behav. Med.* **2017**, *41*, 139–151. [CrossRef]

nutrients MDPI

Article

The Impact of the COVID-19 Pandemic on Public Interest in the Energy Labelling on Restaurant Menus

Areej A. Alkhaldy [1,*], Omar A. Alhumaidan [2], Sarah M. Alkhunein [2], Majid M. Alkhalaf [2], Khlood A. Bookari [3] and Jamila M. Arrish [2]

1 Clinical Nutrition Department, Faculty of Applied Medical Sciences, King Abdulaziz University, P.O. Box 80215, Jeddah 21589, Saudi Arabia
2 National Nutrition Committee (NNC), Saudi Food and Drug Authority (Saudi FDA), P.O. Box 13312, Riyadh 13513, Saudi Arabia
3 Clinical Nutrition Department, Faculty of Applied Medical Sciences, Taibah University, P.O. Box 344, Medinah 42353, Saudi Arabia
* Correspondence: aalkhaldy@kau.edu.sa

Abstract: No study has investigated the effect of the COVID-19 pandemic on the public's interest in using energy labelling on restaurant menus. This study explores the effects of the COVID-19 pandemic on the public interest in using energy labelling on restaurant menus and meal delivery applications and the impact of energy-labelling availability on food choices during the COVID-19 pandemic in Saudi Arabia. An online questionnaire was completed by 1657 participants aged $\geq$ 18 years. Before the COVID-19 pandemic, 32% of customers visited a restaurant 2–4 times/week. However, during the pandemic, 35% of customers visited a restaurant only once per week. There was no difference in interest in reading energy labelling or using meal delivery applications before and during the pandemic. During the COVID-19 pandemic, about 55% of restaurant customers reported that they had noticed energy labelling, with 42% of them being influenced by the energy-labelling information. Regarding energy information on food delivery applications, 40% of customers noticed energy labelling when using the applications, with 33% of them being affected by the energy labelling. Customer interest in reading about energy on restaurant menus during the pandemic did not change significantly from the level of interest before the pandemic. The interest expressed by the public in using the energy labelling was low both before and during the COVID-19 pandemic.

Keywords: menu-labelling policy; nutrition; restaurants; interest

1. Introduction

The emergence of the COVID-19 global pandemic has added several challenges to human lives worldwide, causing remarkable impacts on health, lifestyle, social life, and the economy [1]. In addition, many national and international laws and regulations have been affected, including energy-labelling policies [2,3].

An example of one such impact is the U.S. Food and Drug Administration (FDA) energy-labelling policies. The FDA released a guidance document entitled Temporary Policy Regarding Nutrition Labelling of Standard Menu Items in Chain Restaurants and Similar Retail Food Establishments during the COVID-19 Public Health Emergency to provide temporary flexibility to restaurants and similar food retailers required to provide nutritional information, including calories, on menus and food delivery applications' menu boards [3].

In Saudi Arabia, a menu calorie-labelling policy was announced in August 2018 by the Saudi Food and Drug Authority (SFDA). The policy required all restaurants, including dine-in restaurants, fast-food restaurants, cafes, ice cream shops, fresh-fruit juice shops, pastry shops, higher-educational institutions, and governmental agencies to post caloric information clearly in their menus by the end of December 2018 [4]. Several studies were conducted to evaluate public knowledge, attitudes, views, and practices towards the SFDA's mandatory menu calorie-labelling policy and the impact of the policy on the public's dietary behaviour. Overall, these studies reported adequate knowledge and positive attitudes and practices related to menu caloric labelling in Saudi Arabia. In addition, these studies highlighted the importance of menu caloric labelling as a valuable tool for promoting healthy eating habits and encouraging healthier food selection [5–8]. Unlike the US FDA, the SFDA did not release any temporary guidance to the restaurants in relation to displaying energy labelling during the COVID-19 pandemic.

Many studies have indicated that energy labelling may have a beneficial effect on food choices and is effective in reducing energy order and consumption [9]. During the COVID-19 pandemic, most studies have examined the effect of COVID-19 on eating habits, food intake, physical activities, and lifestyle [9–19], with no available data on the effect of the COVID-19 lockdown on public use and interest in using restaurant-menu caloric labelling worldwide, including in Saudi Arabia. Given the possibility of future public-health pandemics, it is essential to understand the interest of the public in using energy labelling during the COVID-19 public-health pandemic. A better understanding of public perception of energy labelling may provide important information that could be used to raise awareness on how to utilise energy labelling to improve food choices and reduce energy intake during a global health emergency.

It has been reported that negative weight-related behaviours, including unhealthy eating and physical inactivity during the COVID-19 pandemic, were associated with being overweight and obese, which could make these people more susceptible to chronic diseases [20]. To address obesity issues and high food consumption during the COVID-19 pandemic, it is recommended that the energy-labelling legislation be adopted to influence consumer eating choices and reduce the risk of obesity and its complications during any public health emergency.

This study aims to explore the effects of the COVID-19 pandemic on the public interest in using energy labels on restaurant menus and meal delivery applications and whether the availability of energy labelling during the COVID-19 pandemic impacted the public's food choices. The findings of this study will help to understand the public's response toward the caloric labelling of menus in health emergencies. Further, the data of this study could be used and put into consideration when developing the guidance document to instruct the restaurants so that the public can estimate their daily caloric intake and maintain a healthy lifestyle during any public health emergency.

2. Materials and Methods

2.1. Study Design

A cross-sectional questionnaire-based internet study was conducted from August 2020 to January 2021 by researchers from the National Nutrition Committee on the SFDA and researchers from King Abdulaziz University and Taibah University in the Kingdom of Saudi Arabia. The study was approved by the Biomedical Ethics Research Committee at King Abdulaziz University (Jeddah, Saudi Arabia) (Reference No. 469-20). Informed consent was obtained from all study participants.

2.2. Participants and Recruitment

A convenience sample of 1657 Saudi adult participants aged $\geq$ 18 was included. Participants were asked to complete a publicly distributed electronic questionnaire using Google Forms. The electronic questionnaire was distributed using the snowball-sampling method via various social media platforms, such as WhatsApp and Twitter. For WhatsApp, the link to the questionnaire was sent to all contacts in the personal contact lists of this study's authors, including relatives, friends, and neighbors, who were encouraged to send the link to all their contacts within Saudi Arabia. In addition, the link was shared with many WhatsApp social groups. On Twitter, the account of the SFDA tweeted the survey link, and this study's authors requested public figures with high numbers of followers in Saudi Arabia to re-tweet the study questionnaire.

2.3. Sample-Size Calculation

The sample size was calculated using an online calculator (Epi Info) [21] and data obtained from the General Authority for Statistics on 25,192,780 adult residents and citizens $\geq$ 18 years in the Kingdom of Saudi Arabia in 2019 [22]. The calculation showed that a sample of 1083 was adequate, with a 99.9% confidence interval, and hypothesised a 50% frequency-of-outcome factor in the population.

2.4. Study Questionnaire

The questionnaire was composed of three main sections, with 33 questions requiring approximately 15 min to answer. Four Ph.D. experts from nutritional and health backgrounds prepared the survey questions. The questionnaire was subjected to validity, reliability, and internal-consistency testing. Content validity was ascertained by 10 Ph.D. and M.Sc. experts in nutrition and dietetics, and a content validity index of 0.84 was achieved. The questionnaire was revised based on the experts' feedback and comments. Test-retest reliability was carried out and expressed as a correlation coefficient with an average value of 0.79. Internal consistency was assessed by Cronbach's alpha, and values for different subscales ranged between 0.72–0.88. The questionnaire was then revised based on their feedback and comments. The modifications included changing the response options, grouping and breaking down some questions, and including a few more questions. The questionnaire had five sections, as follows: (1) Demographic characteristics of the participants included age, sex, marital status, educational level, work status, household income, field of study, and geographic location. The participants' background information included their perceived overall quality of diet, weight perception, whether they had been diagnosed with COVID-19, their medical status, smoking status, physical-activity level, and if they knew how to calculate the calories. In addition, participants were asked whether they agreed, disagreed, or were neutral about the usefulness of energy labelling during the COVID-19 pandemic. (2) Self-reported weight, body mass index (BMI), and quality of diet before and during the COVID-19 pandemic. (3) Frequency of restaurant visits and use of food delivery applications before and during the COVID-19 pandemic. (4) Interest in energy labelling in restaurant and food-delivery-application menus before and during the COVID-19 pandemic. (5) The public's observation of the menu energy labelling and its impact on participants' food choices during the COVID-19 pandemic.

2.5. Statistical Analysis

Data were checked and entered using standardised entry codes written on a Statistical Package for the Social Sciences (SPSS, Version 23.0; IBM Corp., Armonk, NY, USA) v26 data file. The data were presented as frequencies and proportions. Descriptive statistics were used to examine demographic characteristics and food frequency. To test the difference between two variables, a chi-squared test of independence was used with a p-value < 0.05, denoting statistical significance. Bonferroni correction was used due to multiple comparisons.

3. Results

3.1. Demographics and Background of Participants

Table 1 shows the general characteristics of the study participants (n = 1657). The most frequently reported characteristics of study participants were age between 18 and 24 years (52%), female (81%), single (58%), Saudi (93%), had a bachelor's degree (53%), from a scientific background (38%), student (45%), approximately evenly drawn from across the income categories of no income (29%) or less than 2000 SR (29%), and from the west region (51%) (Table 1).

Table 1. Characteristics and backgrounds of participants (n = 1657).

Variables	n	%
Age (y)		
18–24	854	52
25–39	524	32
40–59	255	15
≥60	24	1
Gender		
Male	314	19
Female	1343	81
Nationality		
Saudi	1539	93
Non-Saudi	118	7
Marital Status		
Single	968	58
Married	574	35
Widower	15	1
Divorced	59	4
Preferred not to say	41	2
Educational Level		
High school or less	646	39
University level	877	53
Higher degrees	134	8
Work Status		
Student	743	45
Employed	510	31
Unemployed	241	15
Retired	163	10
Income (SR) [a]		
No income	482	29
<2000 SR	476	29
2000–4000 SR	206	12
5000–7000 SR	136	8
8000–10000 SR	133	8
>10,000 SR	224	14
Field of Study		
Scientific	615	38
Literature	498	30
Medical	408	25
No specific field	114	7
Region		
Central region	499	30
Northern region	68	4
Southern region	53	3
Eastern region	184	11
Western region	853	51
Diagnosed with COVID-19		
Yes	236	14
No	1421	86

Table 1. *Cont.*

Variables	*n*	%
Medical Diagnosis		
Reported no diseases	1325.6	80
Cardiovascular diseases [b]	83.1	5
Osteoporosis	57.1	3
Diabetes	72.1	4
Anemia	123.1	7
Smoking		
Yes	115	7
No	1497	90
Ex-smoker	45	3
Physical Activity		
Fairly inactive	960	58
Moderately active	538	32
Very active	159	10
Knew How to Calculate Calories		
Yes	671	40
No	652	39
Not sure	334	20
Views About the Helpfulness of Energy Labelling During the COVID-19 Pandemic to Improve Food Choices		
Agree	1241	75
Disagree	22	1
Neutral	235	14
Did not know	159	10

n (%) = Data presented as number and percentage; [a] SR = Saudi Riyal; [b] Included hypertension, hypercholesterolemia, and hypertriglyceridemia.

Regarding participant backgrounds, only 14% of the participants were diagnosed with COVID-19, 80% reported no diseases, 90% were non-smokers, and 58% were reasonably physically active. Approximately 40% knew how to calculate calories and 39% did not. Of the participants, 75% agreed that energy labelling during the COVID-19 pandemic was helpful.

3.2. *Perceived Weight Status, Body Weight, and Quality of Diet*

The majority (47%) of participants reported being within the normal weight range, and 31% reported that their weight was slightly above normal. However, during COVID-19, 33% of the participants reported that their weight had increased, and 32% reported that their weight had not changed. The reported BMI values showed that the participants' BMI did not significantly change before and during the COVID-19 pandemic. Of the participants, 72% reported an adequate-to-healthy diet before the COVID-19 pandemic, with the majority (47%) reporting that the COVID-19 pandemic did not affect their diet. Only 21% of the participants reported that the COVID-19 pandemic worsened their diet (Table 2).

Table 2. Perceived weight status, body weight, and quality of diet before and during the COVID-19 pandemic (n = 1657).

		Answers	n	%
Perceived Weight Status	Before COVID-19 Pandemic (n = 1657)	Below normal	181	11
		Within the normal range	774	47
		Slightly above the normal weight	515	31
		Above the normal weight	123	7
		Not sure	64	4
	During COVID-19 Pandemic (n = 1657)	Yes, gained weight	554	33
		Yes, lost weight	379	23
		No, weight had not changed	528	32
		Not sure	196	12
Body Mass Index *	Before COVID-19 Pandemic (n = 1624)	Underweight	210	13
		Normal	741	46
		Overweight	410	25
		Obese	263	16
	During COVID-19 Pandemic (n = 1622)	Underweight	201	12
		Normal	761	47
		Overweight	378	23
		Obese	282	17
Perceived Diet Quality	Before COVID-19 Pandemic (n = 1657)	Unhealthy	420	25
		Adequate	990	60
		Healthy	194	12
		Did not know	53	3
	During COVID-19 Pandemic (n = 1657)	Yes, it improved it	523	32
		Yes, it made it worse	352	21
		No, it did not affect it	782	47

* Calculated based on self-reported weight and height.

3.3. Restaurant Visits and Use of Food Delivery Applications before and during the COVID-19 Pandemic

Before the COVID-19 pandemic, the percentage of restaurant visits (2–4 times/week) was the most frequent among the study participants (32%). However, during the COVID-19 pandemic, visiting restaurants once per week was the most frequent among the study participants (35%). In addition, there was an increase from 7% to 15% of the participants who reported not visiting a restaurant during the COVID-19 pandemic. However, there was no change in the use of food delivery applications before and during the COVID-19 pandemic (Figure 1).

3.4. Interest in Reading Energy Labelling before and during the COVID-19 Pandemic in Restaurants and Meal Delivery Applications

Table 3 shows participants' interest in reading the energy labelling in restaurants and food delivery applications before and during the COVID-19 pandemic. The majority of participants were either somewhat interested (39%) or not interested (43%) in reading the energy labelling before the COVID-19 pandemic in restaurants. During the COVID-19 pandemic, the percentage of participants who were somewhat interested was 39%, with a slight decrease in the number of participants who were not interested at all (38%). Only 17% of the participants were very interested in reading the energy labelling in restaurants before the COVID-19 pandemic, and 22% were very interested during the COVID-19 pandemic. Age, marital status, educational level, and work status were significant variables associated with the interest in reading the energy labelling before and during the COVID-19 pandemic in restaurants ($p < 0.05$). In particular, single young females with bachelor's degrees were seen to be more interested in reading the energy labelling during the COVID-19 pandemic.

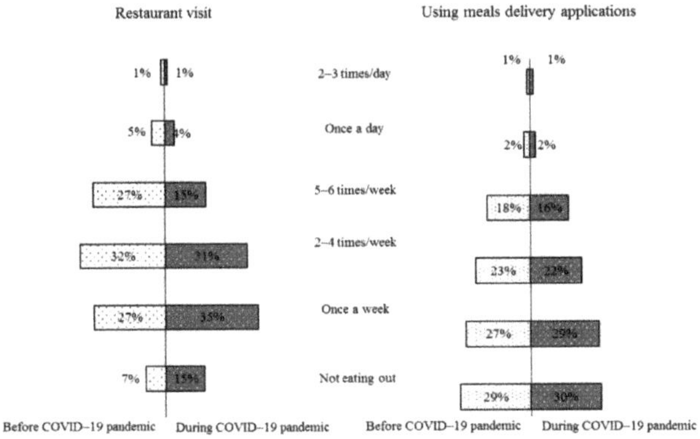

Figure 1. Frequency of restaurant visits and using food delivery applications before and after the COVID-19 pandemic (1657).

Table 3. Interest in reading the energy labelling before and during the COVID-19 pandemic in restaurants and meal delivery applications (n = 1657).

Variables	Restaurant Visits						Food Delivery Applications					
	Before COVID-19			During COVID-19			Before COVID-19			During COVID-19		
	Extremely interested	Little interested	Not interested	Extremely interested	Little interested	Not interested	Extremely interested	Little interested	Not interested	Extremely interested	Little interested	Not interested
Total (%)	17	39	43	22	39	38	17	33	50	20	33	46
Age (y)												
18–24	152 (9)	295 (18)	407 (25)	211 (13)	284 (17)	359 (22)	151 (9)	252 (15)	451 (27)	188 (11)	252 (15)	414 (25)
25–39	93 (6)	233 (14)	198 (12)	103 (6)	245 (15)	176 (11)	81 (5)	207 (12)	236 (14)	101 (6)	205 (12)	218 (13)
40–59	40 (2)	115 (7)	100 (6)	52 (3)	112 (7)	91 (5)	39 (2)	88 (5)	128 (8)	45 (3)	92 (6)	118 (7)
≥60	4 (0)	8 (0)	12 (1)	6 (0)	7 (0)	11 (1)	4 (0)	7 (0)	13 (1)	4 (0)	6 (0)	14 (1)
p-value	0.003			<0.000			0.019			0.011		
Gender												
Male	62 (4)	114 (7)	138 (8)	69 (4)	124 (7)	121 (7)	58 (4)	98 (6)	158 (10)	63 (4)	103 (6)	148 (9)
Female	227 (14)	537 (32)	579 (35)	303 (18)	524 (32)	516 (31)	217 (13)	456 (28)	670 (40)	275 (17)	452 (27)	616 (37)
p-value	0.35			0.973			0.496			0.921		
Marital Status												
Single	181 (11)	350 (21)	437 (26)	242 (15)	346 (21)	380 (23)	182 (11)	303 (18)	483 (29)	215 (13)	302 (18)	451 (27)
Married	88 (5)	260 (16)	226 (14)	106 (6)	267 (16)	201 (12)	77 (5)	218 (13)	279 (17)	96 (6)	222 (13)	256 (15)
Widower	0 (0)	5 (0)	10 (1)	2 (0)	3 (0)	10 (1)	1 (0)	3 (0)	11 (1)	3 (0)	2 (0)	10 (1)
Divorced	10 (1)	24 (1)	25 (2)	13 (1)	19 (1)	27 (2)	10 (1)	17 (1)	32 (2)	13 (1)	18 (1)	28 (2)
Preferred not to say	10 (1)	12 (1)	19 (1)	9 (1)	13 (1)	19 (1)	5 (0)	13 (1)	23 (1)	11 (1)	11 (1)	19 (1)
p-value	0.011			0.001			0.04			0.034		
Educational Level												
High school or less	78 (5)	227 (14)	341 (21)	114 (7)	225 (14)	307 (19)	74 (4)	201 (12)	371 (22)	95 (6)	208 (13)	343 (21)
University level	174 (11)	368 (22)	335 (20)	218 (13)	369 (22)	290 (18)	170 (10)	309 (19)	398 (24)	208 (13)	307 (19)	362 (22)
Higher degrees	37 (2)	56 (3)	41 (2)	40 (2)	54 (3)	40 (2)	31 (2)	44 (3)	59 (4)	35 (2)	40 (2)	59 (4)
p-value	<0.000			<0.000			<0.000			<0.000		
Work Status												
Student	122 (7)	272 (16)	349 (21)	179 (11)	253 (15)	311 (19)	124 (7)	233 (14)	386 (23)	160 (10)	229 (14)	354 (21)
Employed	115 (7)	214 (13)	181 (11)	134 (8)	216 (13)	160 (10)	107 (6)	182 (11)	221 (13)	122 (7)	181 (11)	207 (12)
Unemployed	37 (2)	93 (6)	111 (7)	41 (2)	97 (6)	103 (6)	33 (2)	82 (5)	126 (8)	41 (2)	81 (5)	119 (7)
Retired	15 (1)	72 (4)	76 (5)	18 (1)	82 (5)	63 (4)	11 (1)	57 (3)	95 (6)	15 (1)	64 (4)	84 (5)
p-value	<0.000			< 0.000			< 0.000			< 0.000		

n (%) = Data presented as number and percentage. The p-values for chi-squared tests were calculated among age, gender, marital status, educational level, and work status.

Regarding the use of meal delivery applications, most participants were either somewhat interested (33%) or not interested (50%) in reading the energy labelling both before and during the COVID-19 pandemic. The percentage of participants who were somewhat interested was the same at 33%, with a slight decrease in the number of participants who were not interested at all (46%). Only 17% of the participants were extremely interested before the COVID-19 pandemic, and 20% were extremely interested during the COVID-19 pandemic. Age, marital status, educational level, and work status were significant variables associated with the interest in reading the energy labelling before and during the COVID-19 pandemic when using meal delivery applications ($p < 0.05$). In particular, single young participants with bachelor's degrees were seen to be more interested in reading the energy labelling when using meal delivery applications during the COVID-19 pandemic. In addition, there was a positive association between those interested in reading the energy labelling and their knowledge of how to calculate daily caloric intake (p-value = 0.001).

3.5. The Observation of Energy Labelling and Its Impact on Participants' Food Choices during the COVID-19 Pandemic

Only 17% of participants reported an issue using energy labelling during the COVID-19 pandemic, with 45% reporting no issues and 38% uncertain about their responses. The issues included energy labelling not being displayed next to all meals on the menu, energy labelling not displayed next to some meals on the menu, and no response from the restaurants when asking about energy labelling or inaccurate energy labelling.

During the COVID-19 pandemic, more than half (55%) of participants reported they had noticed energy labelling in the restaurant at their recent visit, with 21% not noticing it and 24% not sure if energy labelling was displayed. About 35% of participants who noticed the energy labelling said that their orders were not affected by the energy labelling, and 23% said they were not sure about their response at that time. However, 42% of the participants who noticed the energy labelling said it affected their order by changing the order (33%), ordering less food (23%), eating less fast food (36%), or choosing another restaurant (8%). Young (age 18–24) ($p < 0.001$), single ($p = 0.022$), and with bachelor's degrees ($p < 0.001$) participants were significantly associated with noticing the menu labelling during restaurant visits.

Regarding the influence of providing energy information on food delivery applications during the COVID-19 pandemic, 39% of the participants reported that they had noticed energy labelling when using food delivery applications, 34% did not notice it, and 27% were unsure if energy labelling was available. About 39% of participants who noticed the energy labelling said it did not affect their order, and 28% were unsure about their responses. However, 33% of the participants who noticed the energy labelling said it affected their experience by changing the order (33%), eating less food (24%), eating less fast food (33%), or choosing another restaurant (9%). Young (age 18–24) ($p < 0.001$), single ($p = 0.009$), and students ($p < 0.001$) participants were significantly associated with noticing the menu labelling on food delivery applications.

4. Discussion

This observational population-based study aimed to learn more about how the COVID-19 pandemic impacted the public's interest into using energy labelling on restaurant menus and food delivery applications across a large, young, and primarily healthy sample of Saudi adults. Our study is one of the first in Saudi Arabia to investigate the immediate impacts of COVID-19 on public interest and behaviour regarding caloric displays following home confinement. There were 1657 respondents, 81% female (n = 1343) and 69% aged 18–30 (n = 1141). The study found that during the COVID-19 pandemic, the participants' interest in reading caloric information displayed on restaurant menus and meal delivery applications had not changed. Most participants claimed that they had little or no interest in energy labelling before or during the COVID-19 pandemic. However, despite their low interest, 75% agreed that energy labelling was helpful in improving individuals' food choices during the COVID-19 pandemic.

This result contrasts with the findings of another major Saudi study that demonstrated considerable gains in closer public attention to nutritional panels and other kinds of food labelling when selecting foods [23]. This could be due to differences in the study population, as the majority of this study's participants were young and healthy, with incomes of about 60% less than 2000 Saudi Riyal. In addition, this study survey was disseminated when the lockdown started to ease. This may have affected the participants' interest in feeling free not to check the energy labelling.

Approximately 40% of the participants claimed that they had enough information to determine their daily caloric needs. This could be attributable to the unchanged BMI results, which showed that 46% of the participants were within the normal BMI range before and during the pandemic. This finding is consistent with the 47% of participants who reported their perceived weight as normal. However, during the COVID-19 pandemic, 33% of participants reported an increased weight. The weight gain could be because about 60% of the participants reported being inactive.

Since the pandemic, people prioritised their health to reduce infection risk by eating healthy [23]. This study found that 60% of the participants reported an adequate diet, with only 12% claiming that their diet was healthy before the pandemic. However, 32% of the participants reported improving their diet. Similar studies in the Netherlands, Poland, Spain, and Qatar reported that participants ate healthier meals during the lockdown [10,24–26]. Health consciousness—a person's health concerns—explains this behaviour [27]. Good behaviours lead to healthy activities in health-conscious people. The pandemic raised health awareness [28]. This feature may be positive and essential in bringing about change, but it may not be enough, especially when changing eating habits [23]. This study found that 47% of the participants did not change their diet during the COVID-19 pandemic. The Transtheoretical Model's phase hypothesis suggests that the coronavirus pandemic may have caused people to contemplate taking action to change their diet [29]. Thus, people may need assistance and educational plans to do so going forward.

During the COVID-19 pandemic curve, people ordering food from restaurants and home delivery applications increased, contributing to a higher daily caloric intake. Recent research in Saudi Arabia showed that 24–50% of diners used restaurant menus' caloric information [5,6,8]. In this study, the percentage of restaurant visits (2–4 times/week) was the most frequent among participants (32%) before the COVID-19 pandemic. However, during the COVID-19 pandemic, visiting a restaurant once per week was the most frequent among 35% of the study participants. In addition, participants who reported not visiting a restaurant during the COVID-19 pandemic increased from 7% to 15%. The majority of participants had visited a restaurant 2–4 times/week prior to the pandemic but this decreased to once per week as a result of the lockdown. In this study, there was no change in use for meal delivery applications before or during the COVID-19 pandemic. Most participants (58%) reported being fairly active, and did not report any significant changes in their BMI during the COVID-19 pandemic. It has been reported that food ordering was significantly associated with weight gain after one month of lockdown among individuals in Malaysia [30]. In contrast, the participants in this study reduced their restaurant meals but did not increase their use of meal delivery services, thus maintaining the same BMI during the pandemic as prior to the pandemic.

During the COVID-19 pandemic, it became necessary to ensure that all restaurants disclosed the energy content of foods in both food menus and meal delivery applications to help people choose healthy food. In this study, 55% noticed energy labelling during their recent restaurant visits, with 42% saying that the energy information changed their meal selections. In addition, 40% of the participants noticed the energy information available on the meal delivery applications, and 33% claimed that the energy labelling changed their orders. For both restaurants and meal delivery applications, the changes in behaviour encouraged participants to change orders, eat less, reduce fast food, or choose a different restaurant.

People's nutritional knowledge is strongly correlated with the type of diet they consume, making education on nutrition one of the many factors necessary to bring about behavioural change. Our findings indicate that less than half of the participants believed they could determine their daily caloric needs. In addition, we found a statistically significant correlation between individuals' interest in energy-labelling information and their knowledge of caloric calculation. These findings are consistent with earlier Saudi [5] and international studies [31,32]. Studies in the West have suggested that educational and nutritional awareness are crucial before making a food choice, increasing people's adherence to the nutritional facts label [33,34]. There are various models of behavioural change, including the Transtheoretical Model [35], the social cognitive theory [36], and the planned behavioural theory [37], that recognise the importance of nutritional-related knowledge, attitudes, and motives of individuals. These "volitional behaviour" models presume a logical connection between a person's thoughts, feelings, intentions, and actions. According to Sapp [38], a "high threshold level of 'how to' and 'awareness' nutrition knowledge" is essential for individuals to decide on healthy choices for eating. On the contrary, erroneous assumptions and inadequate information could create irrational eating habits. Although some studies suggest that nutritional knowledge is irrelevant to eating behaviours, nutritional information is necessary for individuals who adopt a healthy diet and cannot be ignored.

The results of this study should be interpreted with caution due to some limitations. It should be noted that a cross-sectional study can only address relationships between variables; it cannot prove causality. Our study's sociodemographic data may not reflect the proportional structure of the entire Saudi population because we used convenience sampling and social media platforms like Twitter and WhatsApp to disseminate the electronic questionnaire, which may have introduced some minor bias. Convenience sampling is the most efficient and economical way to collect data from a sizeable demographic group [39]. In addition, the online survey was a quick and cost-effective approach to collect self-reported data on participants' knowledge, attention, and behaviour regarding caloric display following home confinement in Saudi Arabia. This was a valuable method for collecting a large number of responses when face-to-face interviews were difficult due to lockdowns. Furthermore, the perception questions did not evaluate whether people knew their energy requirements. However, the study's descriptive findings offer a starting point for exploring how the public reacts to the COVID-19 lockdown regarding their interest in menu caloric labelling.

5. Conclusions

The inconsistent changes of food intake that are reported by individuals may not indicate healthy eating patterns. Still, the pandemic may have instilled a sense of personal responsibility for one's health and boosted people's nutritional interest among the examined Saudi adult population. In Saudi Arabia, the rate of restaurant delivery and takeout has increased, along with the prevalence of obesity and sedentary lifestyles. To encourage a healthy diet and reduce daily calories, it could be helpful to review energy-labelling practices to enhance their use in both normal and health emergency conditions. For those already at a higher risk of obesity, this is of paramount importance, especially during times of crisis and emergency. Education and guidance on energy labelling must be continued to help improve food choices.

Author Contributions: Methodology, A.A.A., O.A.A., S.M.A., M.M.A., K.A.B. and J.M.A.; formal analysis, A.A.A. and O.A.A.; recruitment, A.A.A., O.A.A., S.M.A., M.M.A., K.A.B. and J.M.A.; writing—original draft preparation, A.A.A., O.A.A., S.M.A., M.M.A., K.A.B. and J.M.A.; writing—review and editing, A.A.A.; supervision, A.A.A. and M.M.A.; project administration, A.A.A. and M.M.A. All authors have read and agreed to the published version of the manuscript.

Funding: This research received no external funding.

Institutional Review Board Statement: The study was conducted according to the guidelines of the Declaration of Helsinki and approved by the Unit of the Biomedical Ethics Research Committee at King Abdulaziz University (Jeddah, Saudi Arabia) (reference No. 469-20).

Informed Consent Statement: Informed consent was obtained from all subjects involved in the study.

Data Availability Statement: The data used to support the findings of this study are available from the corresponding author upon request.

Acknowledgments: The authors would like to thank the National Nutrition Committee (NNC) in the Saudi Food and Drug Authority (SFDA) reviewing this study and Budur Bin Obaydan for literature searching.

Conflicts of Interest: The authors declare no conflict of interest.

References

1. Barro, R.J.; Ursúa, J.F.; Weng, J. *The Coronavirus and the Great Influenza Pandemic: Lessons from the "Spanish Flu" for the Coronavirus's Potential Effects on Mortality and Economic Activity*; No. w26866; National Bureau of Economic Research: Cambridge, MS, USA, 2020.
2. Nicola, M.; Alsafi, Z.; Sohrabi, C.; Kerwan, A.; Al-Jabir, A.; Iosifidis, C.; Agha, M.; Agha, R. The socio-economic implications of the coronavirus pandemic (COVID-19): A review. *Int. J. Surg.* **2020**, *78*, 185–193. [CrossRef]
3. Center for Food Safety and Applied Nutrition. Temporary Policy Regarding Nutrition Labeling of Standard Menu Items in Chain Restaurants and Similar Retail Food Establishments During the COVID-19 Public Health Emergency. 2020. Available online: https://www.fda.gov/regulatory-information/search-fda-guidance-documents/temporary-policy-regarding-nutrition-labeling-standard-menu-items-chain-restaurants-and-similar (accessed on 31 October 2022).
4. SFDA Launches Healthy Food Regulation Strategy Tomorrow. Saudi Food and Drug Authority. 26 October 2022. Available online: https://sfda.gov.sa/en/news/1941 (accessed on 31 October 2022).
5. Alkhaldy, A.A.; Taha, D.S.; Alsahafi, S.E.; Naaman, R.K.; Alkhalaf, M.M. Response of the public and restaurant owners to the mandatory menu energy-labelling implementation in restaurants in Saudi Arabia. *Public Health Nutr.* **2020**, *23*, 3435–3447. [CrossRef] [PubMed]
6. Alassaf, H.I.; Alaskar, Y.A.; Alqulaysh, B.F.; Alshehri, M.A.; Alnosian, M.Y.; Alshamrani, A.A.; Al-Tannir, M. Assessment of knowledge, attitudes and practices of Saudi adults about calorie labeling in central Saudi Arabia. *Saudi Med. J.* **2020**, *41*, 296–303. [CrossRef]
7. AlShehri, N.M.; AlMarzooqi, M.A. Consumers' Knowledge, Attitudes, and Practices Toward Calorie Labeling in Riyadh City, Saudi Arabia: A Cross-Sectional Assessment. *Front. Public Health* **2022**, *10*, 893978. [CrossRef]
8. AlAmer, N.A.; AlOmar, R.S.; AlKaltham, S.M.; AlYami, R.S.; AlRashidi, F.N.; AlJrri, M.M.; Wahab, M.M.A. Perceived effect of calorie count display on customers' eating behaviors in food facilities of eastern province, Saudi Arabia: A mixed method study. *J. Multidiscip. Healthc.* **2020**, *13*, 1849. [CrossRef] [PubMed]
9. Littlewood, J.A.; Lourenco, S.V.; Iversen, C.L.; Hansen, G.L. Menu labelling is effective in reducing energy ordered and consumed: A systematic review and meta-analysis of recent studies. *Public Health Nutr.* **2015**, *19*, 2106–2121. [CrossRef]
10. López-Moreno, M.; López, M.; Miguel, M.; Garcés-Rimón, M. Physical and Psychological Effects Related to Food Habits and Lifestyle Changes Derived from Covid-19 Home Confinement in the Spanish Population. *Nutrients* **2020**, *12*, 3445. [CrossRef] [PubMed]
11. Reyes-Olavarría, D.; Latorre-Román, P.Á.; Guzmán-Guzmán, I.P.; Jerez-Mayorga, D.; Caamaño-Navarrete, F.; Delgado-Floody, P. Positive and Negative Changes in Food Habits, Physical Activity Patterns, and Weight Status during COVID-19 Confinement: Associated Factors in the Chilean Population. *Int. J. Environ. Res. Public Health* **2020**, *17*, 5431. [CrossRef]
12. Cheikh Ismail, L.; Osaili, T.M.; Mohamad, M.N.; Al Marzouqi, A.; Jarrar, A.H.; Zampelas, A.; Habib-Mourad, C.; Abu Jamous, D.O.; Ali, H.I.; Al Sabbah, H.; et al. Assessment of eating habits and lifestyle during the coronavirus 2019 pandemic in the Middle East and North Africa region: A cross-sectional study. *Br. J. Nutr.* **2020**, *126*, 757–766. [CrossRef]
13. Almughamis, N.; Alasfour, S.; Mehmood, S. Poor eating habits and predictors of weight gain during the COVID-19 quarantine measures in Kuwait: A cross sectional study. *F1000Research* **2020**, *9*, 914. [CrossRef]
14. Galali, Y. The impact of COVID-19 confinement on the eating habits and lifestyle changes: A cross sectional study. *Food Sci. Nutr.* **2021**, *9*, 2105–2113. [CrossRef]
15. Pérez-Rodrigo, C.; Gianzo Citores, M.; Hervás Bárbara, G.; Ruiz-Litago, F.; Casis Sáenz, L.; Arija, V.; Aranceta-Bartrina, J. Patterns of change in dietary habits and physical activity during lockdown in Spain due to the COVID-19 pandemic. *Nutrients* **2021**, *13*, 300. [CrossRef]
16. Ramos-Padilla, P.; Villavicencio-Barriga, V.; Cárdenas-Quintana, H.; Abril-Merizalde, L.; Solís-Manzano, A.; Carpio-Arias, T. Eating Habits and Sleep Quality during the COVID-19 Pandemic in Adult Population of Ecuador. *Int. J. Environ. Res. Public Health* **2021**, *18*, 3606. [CrossRef]

17. Di Renzo, L.; Gualtieri, P.; Pivari, F.; Soldati, L.; Attinà, A.; Cinelli, G.; Leggeri, C.; Caparello, G.; Barrea, L.; Scerbo, F.; et al. Eating habits and lifestyle changes during COVID-19 lockdown: An Italian survey. *J. Transl. Med.* **2020**, *18*, 229. [CrossRef]
18. Di Renzo, L.; Gualtieri, P.; Cinelli, G.; Bigioni, G.; Soldati, L.; Attinà, A.; Bianco, F.; Caparello, G.; Camodeca, V.; Carrano, E.; et al. Psychological Aspects and Eating Habits during COVID-19 Home Confinement: Results of EHLC-COVID-19 Italian Online Survey. *Nutrients* **2020**, *12*, 2152. [CrossRef]
19. Alhusseini, N.; Alqahtani, A. Covid-19 Pandemic's Impact on Eating Habits in Saudi Arabia. *J. Public Health Res.* **2020**, *9*, jphr-2020. [CrossRef] [PubMed]
20. Robinson, E.; Boyland, E.; Chisholm, A.; Harrold, J.; Maloney, N.G.; Marty, L.; Mead, B.R.; Noonan, R.; Hardman, C.A. Obesity, eating behavior and physical activity during COVID-19 lockdown: A study of UK adults. *Appetite* **2020**, *156*, 104853. [CrossRef] [PubMed]
21. Info™, E. 7.2; Division of Health Informatics & Surveillance (DHIS), Center for Surveillance, Epidemiology & Laboratory Services (CSELS). 2018. Available online: https://www.cdc.gov/csels/dhis/index.html (accessed on 15 June 2020).
22. General Authority for Statistics Kingdom of Saudi Arabia. Population by Age Groups, and Gender Mid Year. Available online: https://www.stats.gov.sa/en/6768 (accessed on 10 June 2020).
23. Alkhalaf, M.M.; Bookari, K.A.; Arrish, J.; Fallata, G.A.; Alhumaidan, O.A.; Alakeel, S.A.; AlBuayjan, N.A.; Alkhunein, S.M.; Bin Obaydan, B.M.; Alshaya, A.A. The Effect of the COVID-19 Pandemic on Saudi Adults' Behavior Regarding Food Literacy and Food Consumption. *Cureus* **2022**, *14*, e27878. [CrossRef] [PubMed]
24. Poelman, M.P.; Gillebaart, M.; Schlinkert, C.; Dijkstra, S.C.; Derksen, E.; Mensink, F.; Hermans, R.C.; Aardening, P.; de Ridder, D.; de Vet, E. Eating behavior and food purchases during the COVID-19 lockdown: A cross-sectional study among adults in the Netherlands. *Appetite* **2020**, *157*, 105002. [CrossRef]
25. Ben Hassen, T.; El Bilali, H.; Allahyari, M.S. Impact of COVID-19 on food behavior and consumption in Qatar. *Sustainability* **2020**, *12*, 6973. [CrossRef]
26. Sidor, A.; Rzymski, P. Dietary Choices and Habits during COVID-19 Lockdown: Experience from Poland. *Nutrients* **2020**, *12*, 1657. [CrossRef] [PubMed]
27. Chen, M.-F.; Lin, N.-P. Incorporation of health consciousness into the technology readiness and acceptance model to predict app download and usage intentions. *Internet Res.* **2018**, *28*, 351–373. [CrossRef]
28. Dutta-Bergman, M.J. Primary Sources of Health Information: Comparisons in the Domain of Health Attitudes, Health Cognitions, and Health Behaviors. *Health Commun.* **2004**, *16*, 273–288. [CrossRef]
29. Shaffer, J.A. Stages-of-change model. In *Encyclopedia of Behavioral Medicine*; Springer: Berlin/Heidelberg, Germany, 2020; pp. 2129–2132.
30. Chen, H.W.J.; Marzo, R.R.; Anton, H.; Abdalqader, M.A.; Rajasekharan, V.; Baobaid, M.F.; Hamzah, H.; Tang, H.C.; Ads, H.O. Dietary Habits, Shopping Behavior and Weight Gain during Covid-19 Pandemic Lockdown among Students in a Private University in Selangor, Malaysia. *J. Public Health Res.* **2021**, *10* (Suppl. S2), jphr-2021. [CrossRef]
31. Bleich, S.N.; Pollack, K.M. The publics' understanding of daily caloric recommendations and their perceptions of calorie posting in chain restaurants. *BMC Public Health* **2010**, *10*, 121. [CrossRef]
32. Radwan, H.; Faroukh, E.M.; Obaid, R.S. Menu labeling implementation in dine-in restaurants: The Public's knowledge, attitude and practices. *Arch. Public Health* **2017**, *75*, 1–7. [CrossRef] [PubMed]
33. Szykman, L.R.; Bloom, P.N.; Levy, A.S. A Proposed Model of the Use of Package Claims and Nutrition Labels. *J. Public Policy Mark.* **1997**, *16*, 228–241. [CrossRef]
34. Petrovici, D.; Fearne, A.; Nayga, R.M.; Drolias, D. Nutritional knowledge, nutritional labels, and health claims on food: A study of supermarket shoppers in the South East of England. *Br. Food J.* **2012**, *114*, 768–783. [CrossRef]
35. Ajzen, I. The theory of planned behavior. *Organ. Behav. Hum. Decis. Process.* **1991**, *50*, 179–211. [CrossRef]
36. Bandura, A. Social Cognitive Theory: An Agentic Perspective. *Asian J. Soc. Psychol.* **1999**, *2*, 21–41. [CrossRef]
37. Prochaska, J.O.; Velicer, W.F. The Transtheoretical Model of Health Behavior Change. *Am. J. Health Promot.* **1997**, *12*, 38–48. [CrossRef] [PubMed]
38. Sapp, S.G. Incomplete knowledge and attitude-behavior inconsistency. *Soc. Behav. Personal. Int. J.* **2002**, *30*, 37–44. [CrossRef]
39. Jager, J.; Putnick, D.L.; Bornstein, M.H., II. More than just convenient: The scientific merits of homogeneous convenience samples. *Monogr. Soc. Res. Child Dev.* **2017**, *82*, 13–30. [CrossRef] [PubMed]

MDPI AG
Grosspeteranlage 5
4052 Basel
Switzerland
Tel.: +41 61 683 77 34

Nutrients Editorial Office
E-mail: nutrients@mdpi.com
www.mdpi.com/journal/nutrients

www.ingramcontent.com/pod-product-compliance
Lightning Source LLC
LaVergne TN
LVHW070446061126
851858LV00011B/170